EMBODIMENT IN QUALITATIVE RESEARCH

Embodiment in Qualitative Research connects critical, interdisciplinary theorizing of embodiment with creative, practical strategies for engaging in embodied qualitative research. Ellingson equips qualitative researchers not only to resist the mind–body split in principle but to infuse their research with the vitality that comes from embracing knowledge production as deeply embedded in sensory experience.

Grounded in poststructuralist, posthumanist, and feminist perspectives, this innovative book synthesizes current interdisciplinary theories and research on embodiment; explores research examples from across the social sciences, education, and allied health; and features embodied ethnographic tales and evocative moments from everyday life for reflexive consideration. Each chapter offers flexible starting points for *doing* embodiment actively throughout every stage of qualitative research. An awareness of, and an active engagement with, issues of embodiment enhances scholars' ability to produce high quality research and enlarges their capacity as public intellectuals to spark positive social change, particularly within marginalized communities. The strategies offered relate to methodologies from across the entire spectrum: from traditional qualitative methods such as grounded theory, critical/theoretical analysis, and discourse analysis, to arts-based research — including performance, autoethnographic narrative, poetry, and documentary film making.

Embodiment in Qualitative Research is designed as a resource book for qualitative researchers who want to explore the latest trends in critical theorizing. The writing style will appeal to researchers who seek a bridge between abstract theorizing and pragmatic strategies for producing outstanding qualitative research, as well as to critical scholars who want to integrate embodied ways of knowing with their theorizing. Graduate (and advanced undergraduate) qualitative methods students and early career researchers, as well as advanced scholars seeking to enrich the scope and texture of their work, will find the text inspiring and engaging.

Laura L. Ellingson, Ph.D., is Professor of Communication and Women's & Gender Studies at Santa Clara University. Her research focuses on qualitative and feminist methodologies, gender within extended and chosen family networks, and communication in health care delivery.

EMBODIMENT IN QUALITATIVE RESEARCH

Laura L. Ellingson

Routledge
Taylor & Francis Group

NEW YORK AND LONDON

First published 2017
by Routledge
2 Park Square, Milton Park, Abingdon, Oxon OX14 4RN

and by Routledge
711 Third Avenue, New York, NY 10017

Routledge is an imprint of the Taylor & Francis Group, an informa business

British Library Cataloguing in Publication Data
A catalogue record for this book is available from the British Library

Library of Congress Cataloging in Publication Data
A catalog record for this book has been requested

ISBN: 978-1-62958-230-6 (hbk)
ISBN: 978-1-62958-231-3 (pbk)
ISBN: 978-1-31510-527-7 (ebk)

Typeset in Bembo
by Swales & Willis Ltd, Exeter, Devon, UK

For Glenn, with love and gratitude

CONTENTS

Brief Table of Contents

Detailed Table of Contents

FIGURES AND TABLES

Figures

Tables

ACKNOWLEDGMENTS

Justice Sandra Day O'Connor, the first woman appointed to the U.S. Supreme Court, and one of my childhood heroes, said, "We don't accomplish anything in this world alone . . . and whatever happens is the result of the whole tapestry of one's life and all the weavings of individual threads from one to another that creates something." This book is the result of tremendous collective effort and many, many threads coming together.

Many thanks to the SCU Provost's Office for the research sabbatical that enabled me to immerse myself deeply in this work and write this book. Thanks to the Department of Women's & Gender Studies, the Department of Communication, and the Office of the Dean of the College of Arts and Sciences for the research and travel support that enabled me to attend conferences where I could test out my ideas about embodiment and receive generous feedback. For their assistance with compiling and checking references, I thank our wonderful Women's & Gender Studies student assistants: Mirelle Raza, Juliet Heid, and Rani Hanstad. My thanks to Helen Otero, Cindie Simms, and Joni Berticevich, administrative goddesses and friends, who generously sort out all the details and bureaucratic paperwork that support my work at SCU. My heartfelt gratitude further goes to graphic artist Elwood Mills of the SCU Media Services Department who prepared the figures for this book.

My academic home is with the members of the Organization for the Study of Communication, Language, and Gender, and our annual conference and our ongoing community are gifts to my heart, soul, and mind. Another part of my heart resides in the National Communication Association's Ethnography Division with Shirley Drew, Lainey Jenks, Bob Krizek, Melanie Mills, Bill Rawlins, and Patty Sotirin, and in my long friendships with colleagues Kathy Propp and Jimmie

Manning. Thank you for making me laugh and making me think. I will always and ever thank the two amazing mentors who started me on my way to being a scholar and have remained my academic aunts, friends, and co-mentors since: Dr. Patrice Buzzanell and Dr. Carolyn Ellis.

In particular, I owe a deep debt of gratitude to a handful of marvelous colleagues who went above and beyond the call of duty, friendship, and collegial relations to help me work out and refine my ideas about doing embodiment and to persevere in writing this book. From the bottom of my heart, thank you for your generosity: Lynn Harter who amazes me and understands the importance of the singular beverage in everyday life, Maggie Quinlan who inspires me and helps me see the world from different angles, Bill Rawlins who is so willing to listen and offer his incredible, invigorating wisdom (and oatmeal!), and Patty Soritin, who not only is brilliant but who calmly walks me through what seem like unsolvable dilemmas until I see a way forward. The wisdom and creativity of these friends is woven throughout this book.

Many thanks to my special sabbatical buddies in my virtual group—Laura, Ulrike, Charissa, and Kim—and to my coffee shop pals, Eileen Elrod and Leslie Gray. Thanks for helping make my sabbatical a lot more productive and a lot more fun than it would have been on my own. Thanks to nursing scholar Rafael Romo who provided the perfect interested-researcher-not-trained-in-critical-theory perspective as my ideas and book evolved, and who was the only neighbor I ever had who wanted to talk about qualitative research methods.

So many other friends and family kindly support my everyday body-self: my parents Jane and Larry, Jim, Brigitte, Zac, Jamie, Mark, Diane, Miette, Eric, Elizabeth, Sam, Nina, Janice, Paul, Alan, Auntie Joan, Kris and Amanda, Marissa, Matt, and especially Genni. Thanks to good friends who add so much to my life: Mike, Justin, Hsin-I, my Dyad, G'linda, Linda and Betty, Diane, Dan, Matt and Mary, Lisa, Connie, Kim, Pauline, Sarah, Casey, Jeanine, and all the members of my wonderful coffee klatch/Dining for Women clan, including our outstanding hosts Barbara and Pat. My deepest gratitude to Eve Solis whose compassion, insight, and amazing spirit has helped me more than I can say, and to Charlie and Jen who make it possible for me to keep on walking.

Many thanks to our 13th Street neighbors who make my daily trip home such a pleasant one. I thank my furry children, Westley and Buttercup, for all the comfort, cuddles, and reminders to play. Finally, I thank my incredible spouse and partner, Glenn, who lives with me and loves me day in and day out. Our intra-actions make me who I am; may we always continue to become together.

INTRODUCTION

Elusive Bodies: The (Dis)Embodiment of Research

> The body is our method, our subject, our means of making meaning, representing, and performing.
>
> *(Perry & Medina, 2011, p. 63)*

Researchers begin with the body. Although some researchers remain unconscious of it (or even deny it), embodiment is an integral aspect of all research processes, including qualitative, quantitative, and critical inquiry. For example, researchers often select participants to interview based upon physical characteristics (e.g., race, gender, age) and bodily experiences (e.g., living with multiple sclerosis, tattooing). In ethnographic sites, bodies encounter each other as warm, material manifestations of ourselves; bodies do not wait quietly outside while a core ethnographic self interacts with the equally disembodied minds of participants. Nor do free-floating brains analyze data independently of the fingers (or voices or eyes using adaptive technologies) that turn pages, press keys, and wield pens. Likewise, representation is an embodied act; researchers write or type (or draw or paint or photograph or dance) and discover new meanings even as we move across the page, stage, canvas, or screen.

Yet despite more than three decades of discourse among qualitative, feminist, postmodern, poststructuralist, critical race, postcolonialist, and other critical researchers about the centrality of embodiment to research and sense making, many qualitative researchers still do not know *how* to deliberately embody their practices in ways that make bodies a meaningful presence in their research. I address this opportunity by framing qualitative research as an always already *embodied communicative process*. In this book, I connect theory and epistemology directly to practical strategies for engaging embodiment in research practices.

The goal of this book is to invite and equip qualitative researchers to infuse their research with the vitality of embodiment. This book synthesizes current scholarship on embodiment, explores research exemplars from a variety of disciplines, and features a series of embodied, ethnographic tales from my own research and autoethnographic tales from my everyday embodied life. Rather than rules, formulas, or recipes, I advance creative, flexible starting points for *doing* embodiment reflexively throughout every stage of qualitative research. An awareness of and an active engagement with issues of embodiment enhances scholars' ability to produce high quality research and enlarges our capacity as public intellectuals to spark positive social change, particularly with underserved and marginalized communities.

Defining Embodiment

Poststructuralists (e.g., Manning, 2013) reject the metaphor of the body as a container of the self and instead theorize the body "as a material and visceral set of biological components and functions" (Ash & Gallacher, 2015, p. 69) and "a site through which a situated sense of self is experienced" (Sekimoto, 2012, p. 232). Embodiment refers to

> bodies as whole experiential beings in motion, both inscribed and inscribing subjectivities. That is, the experiential body is both a representation of self (a "text") as well as a mode of creation in progress (a "tool"). In addition, embodiment is a state that is contingent upon the environment and the context of the body.
>
> *(Perry & Medina, 2011, p. 63)*

Beliefs about the relationship between the mind and body point to a fundamental tension surrounding what one means when one speaks of embodiment or "the body." On the one hand, philosophers and scholars construct a notion of the body as a material entity whose potential meanings are constituted and circumscribed by culture(s) through particular discursive systems that privilege certain sets of norms and values (e.g., global capitalism, the U.S. prison industrial complex, religious doctrines) (Gergen, 1994). On the other hand, bodies can also be understood as containing our essential qualities (e.g., emotions, gut instincts, physical characteristics) and material being. That is, cultural meanings certainly vary and have dramatic impact on how we come to interpret bodies, but we cannot completely disassociate such meanings from the concrete, physical reality of the body as a lived entity (e.g., Marshall, 1999). Thus "the body . . . is simultaneously physical and affective, social and individual, produced and producing, reproductive and innovative" (Jones & Woglom, 2015, p. 116). Researchers attend to bodies in order "to find the particularities in how minded bodies and worlds fit together" (Pitts-Taylor, 2015, p. 23). Body studies scholars also usefully challenge the notion that humans are the natural center of meaning with "the human as the measure and measurer of all things" (Ash & Gallacher, 2015, p. 69). Human embodiment is mutually constitutive with the world around us and that understanding can inform qualitative research practices.

Unapologetically Eclectic

I practice my feminist qualitative scholarship betwixt and between spaces dedicated to the focused pursuit of a single theoretical, methodological, or disciplinary sphere; I am a "theoretical fence sitter" (Avner, Bridel, Eales, Glenn, Walker, & Peers, 2014, p. 55) who engages in openly "promiscuous analysis" (Childers, 2014). Embodiment necessitates an interdisciplinary approach (Pitts-Taylor, 2015), and I have incorporated research and theory from not only my home disciplines of communication studies and gender studies, but also education, sociology, anthropology, nursing, critical geography, and an array of interdisciplinary fields—body studies, disability studies, women's studies, fat studies, queer studies, mad studies, science studies. Burke (1973) urged scholars to "use all there is to use" (p. 23) in our work, and I creatively pull threads across disciplines, paradigms, and methodologies (see Ellingson, 2009a). Border work intrigues me. I do not claim to have equally incorporated all the various perspectives in embodiment theorizing as it pertains to qualitative research, let alone embodiment theory and application more broadly. However, I value pragmatism in methodology (and ethics, which I address in Chapter 2), and I use the tools that get the job done.

In casting a wide net for concepts, ideas, and tools for doing embodiment, I may have profaned (or omitted) what is sacred ground for some readers. I readily acknowledge that my choices are grounded in both my feminist political commitments and my background in researching and teaching health communication and gendered communication. Moreover, I have had to omit far more material than I included, skimming over very important details and distinctions in order to bend concepts toward this volume's purpose. This eclectic approach to embodiment is focused on my goal of offering pragmatic inroads for those would like to engage in more deliberate embodiment of their qualitative research processes. I want to accomplish this purpose in a manner that is accessible to scholars schooled in a wide range of methodologies, that is, people who do not necessarily agree with each other on many aspects of methods (specific practices of research such as writing fieldnotes or conducting an interview), epistemology (ways of knowing and establishing what counts as valid evidence), ontology (ways of understanding the nature of being and reality), and axiology (values that form the basis of ethical choices in research). I am engaged in an act of synthesis, translation, connection, and creativity. Throughout, I have provided cites for sources where those interested in particular theories and topics can read in more depth.

My perspective is influenced by myriad theoretical perspectives: see the helpful table provided by Perry and Medina (2015, p. 4) which categorizes theoretical perspectives on embodiment (naturalistic, semiotic, phenomenological, poststructuralist and affect theory, social theory/feminism, Foucaultian, and posthumanist). Further I am indebted to Pink's (2009) articulation of sensory ethnography, who draws upon philosophers of embodiment, including Merleau-Ponty (1962, 1968), Gibson (1966, 1979), and Deleuze and Guattari (1987). Additionally, I benefited from methodologists who draw on Deleuzian concepts (MacLure, 2013; St. Pierre & Jackson, 2014), new materialist feminists (e.g., Grosz, 1994; Hekman, 2010) and posthumanists (Barad, 2007). I also reference the burgeoning field of neurofeminism (Schmitz & Höppner, 2014).

Further, I refuse the implicit, and often explicit, resistance to providing concrete instructions for researchers based upon the epistemological or methodological arguments put forth in theoretical essays. This commonly expressed refusal (even disdain) is expressed as an unwillingness to be formulaic or deterministic. For example, one excellent methodological essay[1] explained that "we did not provide a step-by-step process for engaging in mapping activity (an idea that would seem ludicrous to Deleuze and Guattari, based as it would be in modernist, humanist principles)" (Martin & Kamberelis, 2013, p. 677). Likewise, an intriguing essay on intersectionality and feminist inquiry explained that the author did "not provide written-in-stone guidelines," nor "a kind of feminist methodology to fit all kinds of feminist research," and certainly did "not produce a normative straitjacket for monitoring feminist inquiry in search of the 'correct line'" (Davis, 2008, p. 79).

I grant that any attempt to offer detailed instructions for doing research is necessarily modernist, but I submit that it need not be *merely* modernist. Instead, methodologists can describe processes that include flexible possibilities, as jumping off points rather than authoritative endings. Moreover, the refusal to offer specific steps may be elitist; this practice makes it very difficult for eager researchers who already have a grasp of qualitative practices to experiment with new techniques, theories, and goals. Some methodologists may argue that researchers must read and understand a considerable amount of the complex critical theory in which methods are steeped before they attempt methods, but I respectfully disagree. Essays that feature feminist, postmodern, poststructuralist, posthumanist, postcolonial, critical race, queer, crip/disability, and other forms of critical theorizing often are extraordinarily dense and require familiarity with histories of theoretical conversation and extensive vocabularies. Such theoretical forays need to be complex and utilize many concepts in order to parse the subtle distinctions and articulate nuanced possibilities for making sense of the world. However, such theoretical discussions also present a barrier to accessibility and application of useful ideas.

Despite my extensive background in qualitative methods and critical theory, after reading many fascinating accounts of theoretical and methodological forays, I had no idea how to go about engaging in practices for which the authors had provided me theoretical background and exemplars. Students (and the rest of us) grasp theoretical understandings and methodological practices through iterative processes of reading, engaging, and reflecting, not in a linear fashion. So I provide some instructions here, as I did in my articulation of a crystallization framework for multimethod/multigenre qualitative research (Ellingson, 2009a), so that researchers have a place to begin to weave and wander their own paths.

Another aspect of my eclectic approach is my embodied standpoint. As with my other scholarship, this "text emerges from the researcher's bodily standpoint as she is continually recognizing and interpreting the residue traces of culture inscribed upon her hide from interacting with others in contexts" (Spry, 2001, p. 711). To share some of the most salient aspects of my embodiment, I am a 28-year survivor of osteosarcoma (bone cancer): my right leg was amputated above the knee following two decades of reconstructive, "limb-salvaging" surgeries. I rely on a computerized leg prosthesis for daily mobility and cope with chronic phantom limb pain. My troublesome body demands continual attention and accommodation, making it impossible to ignore the ways in which embodiment necessarily affects (and is affected by) my research processes, relationships with participants, and perspectives on knowledge construction (Ellingson, 1998, 2005, 2006, 2012). Yet my body is not unique in its relevance; *all* researchers' bodies play important roles in producing (all types of) qualitative research. Further, I am a female body who benefits from white, heterosexual, and middle class privileges as they intersect with sexism and ableism. I am in my late 40s and am in a committed heterosexual relationship. I strive to be reflexive about how my standpoint sparks, constitutes, and circumscribes my meaning-making processes (Harding, 1991). I return to the intersectionality of identities throughout the book in an effort to minimize essentializing of bodies into static, totalizing categories.

The Mind-Body Split and the Erasure of Bodies in Research

The privileging of the mind over the body is deeply engrained in Western cultures and hence within conventional research methodologies. The mind–body separation posits "a clear division between mind, equated with self, experienced as proactive and unthreatening, and body, experienced as potentially troublesome" (Marshall, 1999, p. 71). Under the legacy of Descartes' philosophy, rationality dictates that the (higher) mind-self should seek to control its body-property, preferably to the point of rendering it absent, or at least irrelevant, to any knowledge project (often referred to as "Cartesian dualism"). Further, Western cultures reinforce a dichotomous view of embodied gender in which masculinity is associated with the rational mind and knowledge production and femininity with the (reproductive) body and the subjectivity of emotion (e.g., Haraway, 1988). Cartesian and Kantian reasoning reinforces dichotomies that "have made it extremely difficult to find a place in our views of human meaning and rationality for structures of imagination" that would help us to think more holistically about both the limits of emotional responses and the misuses of rational logics (Johnson, 1987, p. xxix). Such a mind–body split renders bodily knowledge oxymoronic; indeed, "it is as if 'facts' come out of our heads, and 'fictions' out of our bodies" (Simmonds, 1999, p. 52).

Typically, social science researchers conduct and represent research as though knowledge were produced without unruly bodies involved. The performance of "disembodied researcher" has been repeated for so long that it functions as a set of naturalized norms that privilege a masculinist rationality as the only legitimate form of knowledge, accorded only to those with sufficient social privilege to deny their feminine unruliness (Ellingson, 2006). Leaving (our own and others') bodies unmarked in our reports and other representations is the privilege of the powerful. Journal articles and other research accounts largely reflect social science norms that frame the researcher's personality, body, and other sources of subjectivity as irrelevant. Disembodied prose appears to come from nowhere, implying a disembodied author (Haraway, 1988). Researchers have used the power of academic discourse to define their bodies as essentially irrelevant to the production of knowledge (Denzin, 1997). When researchers' bodies remain unmarked in our accounts, they reinscribe the power of scholars to speak without reflexive consideration of their positionality. Awareness of the embodiment of knowledge production has grown "in the wake of the various 'turns' that have convulsed the humanities and social sciences: poststructuralist, postmodernist, deconstructive, Deleuzian, performative, affective, material feminist" (Koro-Ljungberg & MacLure, 2013, p. 219), as well as narrative (Ellis & Bochner, 2000), postcolonialist (Mohanty, 1988), and posthumanist (Barad, 2007).

Less powerful researcher bodies do not have the privilege of disowning their unruly physicality; scholars with queer, disabled, nonwhite, Third World/Global South, and otherwise marked bodies encounter resistance to claims of disembodied prose and the privilege of objectivity (e.g., Brown & Boardman, 2010; Sharma, Reimer-Kirkham, & Cochrane, 2009; Simmonds, 1999). Likewise, often researchers render participants' embodied experiences marginal to our research findings or else unproblematically offer their bodies up as evidence by failing to problematize them at all, thus essentializing bodily markers such as race, gender, or sexual orientation as constitutive of meanings about members of marginalized groups.

An alternative (albeit still inevitably imbued with power) is to embrace embodiment as a resource for qualitative research, to position it as an unbounded set of flexible embodied practices—cognitive, emotional, physical, reflexive, engaged—that researchers can *do*. Drawing upon Merleau-Ponty, Barnacle (2009) suggests that we conceive of ourselves as "body-subjects" whose embodied senses provide the basis for us to learn and internalize cultural rules and norms, forming the basis for embodied formal learning. Rational thought cannot be split off from this embodied knowledge production process: "any adequate account of meaning and rationality must give a central place to embodied and imaginative structures of understanding by which we grasp our world" (Johnson, 1987, p. xiii). I embrace Wacquant's (2009) openness to embodiment and "the distinctive possibilities and virtues of a carnal sociology" in which "the social agent is a suffering animal, a being of flesh and blood, nerves and viscera, inhabited by passions and endowed with embodied knowledges and skills" (p. 209). In other words, I posit that qualitative and critical researchers can *do* embodiment in all aspects of research, and that embodiment is integral to both research processes and products. Such embrace of embodiment or carnality does not require a rejection of attention to discourse. Barad (2007), who argues that bodies are too easily dismissed in postmodern theorizing that centers discourses as the focus of inquiry, brings the body back in, not to the exclusion of discourse, but rather by bringing the discursive and the material together for examination. For Barad, humans, nonhuman material objects, and discourse all work together to create different "becomings," rather than static meanings or essences. Attention to embodiment requires holding cultural discourses of, say, disability and ableism, as mutually constitutive with material bodies framed by discourses as able/disabled.

Qualitative researchers willing to experiment with doing embodiment actively in their research (on any topic) will find several benefits to doing so. First, embodied research enables scholars and practitioners to learn about topics that are otherwise unknowable. While one can certainly learn about experiences such as pregnancy by studying popular culture discourse on pregnancy—the representation of pregnancy in movies, television, novels, self-help books—a focus on the embodied experiences of pregnant women sheds light on the topic in a way that highlights the corporeal nature of the condition (Walsh, 2010). For example, pregnant and laboring women explain their embodied sense of being pushed and pulled between competing discourses of medicalized and natural childbirth (Parry, 2006; Turner, 2002, 2004).

Second, explicit attention to embodiment can open new possibilities for analysis and representation in qualitative research. Embodiment has always been implicit in all forms of analysis and representation, even thought it has often been merely bracketed in attempts to eliminate "bias." With attention to embodiment, those coding data could attend to sensory terms, descriptions of body parts or bodily functions, or implications of bodily knowledge. Creative forms of representation of research findings could tap into multiple senses in multimedia formats to go beyond writing to include images, smells, sounds, textures, or other sensory cues. Those writing autoethnographic or ethnographic narratives could attend to details of their characters' bodies. For example, Gibbs (2014) discusses the possibilities for exploring the "more-than-human world" of "Siteworks," an art–science collaborative research project on the Shoalhaven River in southwestern Australia and the "politics of belonging" (p. 208) for people and for elements of the natural world (i.e., water, plants, land).

Third, researchers can enhance the postmodern validity of their studies by incorporating attention to embodiment. As I and many others have explored previously (Ellingson, 2009a), contrasting genres of research representation can generate rich possibilities for validating research in innovative ways. Explicit attention to embodiment manifests a dimension that is often missing from published accounts of research, and at times included in problematic or overly simplistic ways in other formats for sharing research findings and implications with practitioners and general audiences. For example, Gibbs (2014), a cultural geographer and environmental social scientist, explained collaborating with artists to produce their art-science, multi-disciplinary installation, highlighting

> the extent to which my artist collaborators use the body and senses to explore and learn, as well as create and communicate. . . . through various interaction with the site, by gathering, digging, walking, rowing, lugging, listening. We engaged bodily with dirt, documents, river water, the homestead, echoes, cattle, people, weather, cameras, projectors, bird song.
>
> *(Gibbs, 2014, p. 219)*

Adding details of embodiment to complement other types of sense making adds another facet of the crystal through which researchers can understand complex topics, relationships, and identities. At the same time, embodiment and other facets highlight the situated partiality of *all* knowledge claims—not as weakness, error, or "bias," but as inevitable, intriguing sites that point to the always *becoming* status of what researchers think we know.

Fourth, engaging multiple audiences through public intellectualism may be facilitated when embodied elements of research processes and outcomes enable audiences beyond academia to respond viscerally and with immediacy to messages researchers share. Ideas sparked by research can soak into people's being and touch their guts, hearts, hands, and other sensory points as well as their thoughts. A performance that combines physical movement, emotional appeals, and information, for example, could lure the attention of reluctant or apathetic audiences. A written account could be enlivened with photos, participants' drawings, or other visual material that helps illustrate the material elements of a particular lifeworld. For example, Harter, a feminist, qualitative communication studies scholar, explored narrative medicine and patient-centered communication at a national cancer center's sarcoma program through both research articles (e.g., Harter, Patterson, & Gerbensky-Kerber, 2010) and a widely distributed documentary film (Harter & Hayward, 2010). The documentary added sound and images, sharing participants' embodied voices, their homes, and material and discursive elements of health care settings to present a rich tapestry of personal stories of bodies in crisis and those who care for them.

The theorizing and practices of doing embodiment presented in this book form a flexible and dynamic approach that can be incorporated into virtually any qualitative research project, from postpositivist, structured analyses of large data sets all the way through autoethnographic and performative representations. Moreover, embodied research practices help to open possibilities for research projects to span the paradigmatic and representational continuums with multiple methods of analysis and multiple genres of representation of data.

Organization of the Book

This book offers a mix of theoretical, practical, and imaginative approaches to understanding embodiment within qualitative research projects and mixed methods projects that include qualitative methods. Each chapter begins with an ethnographic or autoethnographic narrative from one of my research projects or from my life. Then I synthesize writings by other theorists and methodologists on topics relevant to the chapter focus. The third section of each chapter offers hands-on, practical suggestions for how to meaningfully engage embodiment for researchers and our research processes, participants, and accounts of research.

Following this introduction, I offer an overview of theoretical concepts that constitute *a* model (not *the* model) of doing embodiment in qualitative research. These concepts are embedded in a variety of overlapping theoretical legacies—postmodern, feminist, poststructuralist, new materialist, posthumanist, critical race theory—that offer great insights for opening up postpositivist and interpretive qualitative research to incorporate critical insights. These concepts will be referenced throughout the remainder of the book. Chapter 2 addresses designing and planning for research projects, including attention to the ethics of embodiment, and Chapter 3 overviews bodies studies research with an explicit focus on embodied identity and experiences. I then turn to exploring embodied practices in ethnographic methods and interviewing methods, as the two main forms of collecting data face to face (and via technology), in Chapters 4 and 5 respectively. Next, I offer a chapter on constructing embodied data through recordings, field-notes, interview transcripts, and participant-generated data. Analysis of data is the next topic, and in Chapter 7 I discuss the ways in which all types of data analysis involve researchers' and participants' bodies in complex and exciting ways. Issues of representation are addressed in Chapter 8, covering a full range of genres, from autoethnography, performance, and arts-based research, all the way through to the other end of the methodological continuum where social scientific forms of structured analysis are represented in more conventional research report forms that also can expand to include embodiment in creative and innovative ways. Finally, the Postscript draws together common threads that weave throughout the book and invites readers to do embodiment in their own ways, using particular practices that best serve their embodied, emplaced research projects.

Note

1 I do not intend to demonize these excellent articles by accomplished scholars, only to borrow their language as illustrations of the resistance of theorists to articulating specific practices for qualitative researchers.

1

COMING TO TERMS

Embodying Qualitative Research

Doing Legwork: Inspiring Bodies

I lean forward eagerly, my spine straightens, and my eyes focus intently on the database link on my laptop screen. The title entices me: "Promiscuous Analysis in Qualitative Research." A little shiver of delight passes through me as I silently mouth the words "promiscuous analysis," savoring the pairing of a standard methodological term with a naughty, sexually charged term in what I assume will be a feminist reframing of boundary crossing.

I slide my left index finger across my laptop's cool, smooth track pad and click gently on the glowing blue link to access the article. Quickly scanning the abstract, I feel my gut tighten with excitement and recognition of common ground with the author, Sara Childers, a scholar with whom I am not familiar. My mouth forms a bright grin as I gleefully click the button to download the PDF of the article. I think of the old days when I was a graduate student, and I used to wander the library stacks looking for bound volumes of journals, flipping through to see whether a particular article was worth the hassle and expense of photocopying and shudder at the memory.

I've become addicted to the adrenaline rush of finding more and more journal articles online—the perfect procrastination strategy for avoiding writing—and I have downloaded hundreds of them, zealously inhaling embodiment theorizing in a wild, haphazard assemblage. As I read Childers' (2014) article, I find myself nodding, dragging my finger along passage after passage and selecting the "highlight text" function of my laptop's e-reader application. This electronic marking of a PDF pales in comparison to the tactile satisfaction that comes from the application of purple pen to printed paper, and I sigh wistfully—but only for a moment. I maintain one hardcopy file of my favorite printed articles that

I previously underlined and annotated passionately, but I surrendered to the sheer volume of articles and made the move to electronic retrieval, storage, and interactivity a few years ago. Childers cites an intriguing book chapter from MacLure (2013), and I utter a long thoughtful, "Hmmm," to myself, intrigued. A series of clicks efficiently orders the chapter from my university library's interlibrary loan service, and I smile in anticipation and then nod, pleased.

As my reading continues, my body-self bounces in my seat with joy several times when I read provocative titles, gasps over intriguing concepts, and hums through fascinating analyses of aspects of body-self (e.g., Harris, 2015; Keilty, 2016; Law, 2004). I chuckle self-deprecatingly as I think of how many times people have asked me about my sabbatical writing project. A habitually rapid talker, I regularly find myself excitedly racing through explanations of my book manuscript, waving my hands, and beaming at amused colleagues. I open Birk's (2013) wrenching tale (and thoughtful analysis) of enduring not only years of chronic pain but also skepticism and even dismissal of her suffering from health care providers (and many other people); it resonates so deeply with my own struggle with chronic phantom limb pain that as I read I felt a deep ache in my heart, nausea in my stomach, and a simultaneous relaxing of my shoulders and my facial muscles as the sensation of recognition and affirmation settled over my body, and I repeat in a soft whisper, "Me, too. Me, too. Oh me, too!" (see Mairs, 1997). Sifting through my amassed wealth of concepts, exemplars, theory, narratives, poetry, photography, video, and analyses energizes me through the rest of the afternoon, as I continue to slog through the massive process of (re)organizing and (re)writing my book.

"I'm such a methods geek!" I chuckle to myself. Yet I acknowledge that my vivid embodied responses point to embodied knowing—my gut, lungs, spine, and muscles signal wildly that I am on to something big.

Theorizing Embodiment

I organized and distilled the bodies of theory and research into seven key concepts that will be used throughout this book to discuss doing embodiment as an active engagement with reflexivity, sensuousness, and methods. I lay the concepts out here first to avoid excessive redundancy later and to set the stage for discussions of data collection, analysis, and representation: doing bodies, sensorium, embodied knowing, sticky web of culture, intersubjectivity, actants, and flux.

Doing Bodies

The English language and Cartesian philosophy render the body the possession of the self, as equated with the mind. An alternative approach integrates body,

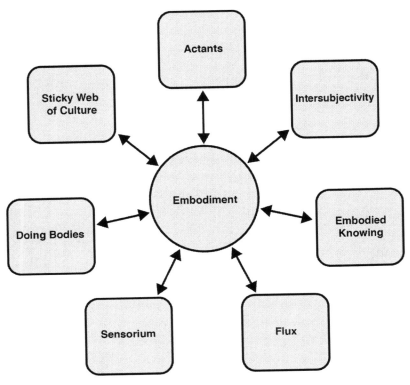

FIGURE 1.1 Conceptualizing Embodiment.
These theoretical concepts help illuminate embodiment in qualitative research practices.

mind, and spirit: "we do not *have* bodies, we *are* our bodies" (Trinh, 1999, p. 258, original emphasis). We enact our body-selves in everyday life, and as such, we *do* our bodies. As Butler (1997) suggests:

> The body is not merely matter but a continual and incessant *materializing* of possibilities. One is not simply a body, but, in some very key sense, one does one's body and, indeed, one does one's body differently from [others]. . . the body is always an embodying of possibilities both conditioned and circumscribed by historical convention.
>
> *(p. 404; original emphasis)*

Thus the body is understood to not exist as a possession of the mind, nor does the self exist in the mind and exert ownership, despite the linguistic convention, "my body," which suggests just such an arrangement. This linguistic shift grounds researchers' attention in embodied, contextualized interactions accomplished in lived spaces, rather than in abstract meanings. Throughout this book, I use the term "body-self" to emphasize the body as self and as everyday performance, and to resist the mind(self)—body dichotomy.

People do our bodies in relation to others as they do bodies, never in isolation, such that "being embodied is always already mediated by our continual interactions with other human and nonhuman bodies" (Weiss, 1999, p. 5). I address non-human bodies (actants) below. For now, Weiss' point that we do not experience our bodies in isolation but in relationship to others is essential. While individuals have sensory experiences of our own, we learn to make sense of them in rela-tion to others' bodies. Hence, we learn to do our bodies in very specific, often highly stylized ways (Butler, 1997). And the ways in which we do bodies shapes the ongoing development of our brains; neurofeminist research establishes "the inseparable entanglements between the development of biological matter and social influences" (Schmitz & Höppner, 2014, pp. 1–2).

Practice theory helps shine a light on *doing* bodies in all sorts of contexts (Hopwood, 2013). Practices are "embodied, materially mediated arrays of human activity centrally organised around shared practical understandings" (Schatzki, 2001, p. 2). Thus the body becomes not merely an instrument for (verbal and nonverbal) communication with others but also the material self that is con-structed through interaction with other bodies and material objects. For example, dialysis care providers communicate as material body-selves within a clinic for patients with kidney failure that also constitutes the health care providers' work-place (Ellingson, 2015). A dialysis technician's practice of offering his arm, a few kind words, a smile, and a gentle pat on the arm to a frail patient while complet-ing biomedical treatment and measurements can be understood within the realm of appropriate professional practice when communicating with a patient. As such, these and other (verbal and nonverbal) communication choices form part of a horizontal web (or mesh, or nexus, or net) of practices that hang together. The practices in which they (bodily) engage come to constitute situated meanings of dialysis care giving and care receiving. Hence participants included bodies with kidney failure, bodies trained to provide dialysis treatment, and bodies tracked as part of a health care organization that serves some and employs others. Moreover, I did my body as an ethnographer in that space, shaping my own body-self and those of the staff, patients, and visitors. In some ways my body conformed to the ways in which staff did bodies—such as wearing a white lab coat over my clothes—while in other ways I did not fit with them, notably in that my body did not engage patients' bodies directly in the process of medical treatment via needles, tubing, and dialysis machines, only in conversation and minor tasks. Patient bodies were done through submission to the invasive treatment, which led most patients to nap, while others watched TV or sought conversation with nearby patients, staff, or myself. Thus doing bodies as a concept shifts the focus of analysis from "having a body" to "being a body" in specific circumstances such that identity goes from fixed to practices that constitute selves.

Sensorium

By far, sight and hearing are privileged culturally and methodologically as the most critical for interfacing with society, yet the other senses also are integral components of individuals' *sensorium*, that is, their "way of coordinating all the body's perceptual and proprioceptive signals as well as the changing sensory envelope of the self" (Jones, Arning, & Farver, 2006, p. 8). Both researcher and participants embody varying physical capacities for perception and participate in lifelong socialization to attend to, and make sense of, sensory perceptions in culturally specific ways. Moreover, we tend not to talk about, or even recognize, other aspects of our senses, unless a failure makes the generally ignored function become suddenly apparent, or make a "dys-appearance" (Leder, 1990).

Neuroscience, social sciences, and philosophy have all contributed to a robust understanding of human senses that goes far beyond the basic conceptualization of five senses. People raised in Western cultures are generally taught that we have five separate senses—sight, hearing, touch, taste, and smell, each of which is associated with a specific part of the body, respectively, eyes, ears, skin (primarily hands), mouth, and nose. Current research suggests, however, that this way of understanding human senses as discrete, parallel processes is inaccurate. The senses do not directly correspond each to a specific organ, with all perception then traced to individual, separate senses, which are captured in a given medium – sight in written word, hearing in music, and so on (Pink, 2009). Our perception both shapes and is shaped by the way in which we talk about senses as separate from one another. "[B]ecause one category is never enough to express exactly what we have actually experienced, the illusion of the 'separate' senses operating *in relation to* each other is maintained" (Pink, 2011, p. 266; original emphasis). Instead, we embody a complex synthesis of sensory perceptions and sense making, a more holistic sensing process performed by the body. Phenomenologist Merleau-Ponty addressed this holistic process decades ago, positing that bodies are not "a collection of adjacent organs but a synergic system, all of the functions of which are exercised and linked together in the general action of being in the world" (Merleau-Ponty, 1962, p. 234) (see also Ingold, 2000; Pink, 2011). The human nervous system is incredibly complex, involving not just all of the senses, but for each sense, multiple types of nerve receptors; for example, "[touch] is itself a combination of data primarily from receptors responding to pressure (mechanoreceptors), temperature (thermoreceptors) and pain (nociceptors)" (Paterson, 2009, p. 770).

Beyond the usual five senses, the "somatic senses" refers to how we experience external and internal sensations that are complex and have to do with balance, muscle tension, and movement: proprioception perceives the positions of one's body in space, while kinesthesia captures a sense of bodily movement, and the vestibular sense gives us a sense of balance (Paterson, 2009). The somatic senses also demonstrate that one cannot separate sensations inside and outside the body, resisting "any neat distinction between interoception and exteroception in the ongoing nature of somatic experiences, and consequently troubles the notion of the haptic as clearly delimited within an individuated body" (Paterson, 2009, p. 780). Thus bodies are deeply enmeshed in their environment—temperature, topography, and so on—including other people's bodies that come into contact either directly or indirectly eliciting embodied responses—e.g., tears, shaking in fear, squirming with discomfort—and non human objects, including other species but also video screens, paintings, or architecture that arouse embodied responses.

I have become aware of the somatic senses firsthand through having a series of reconstructive surgeries on my right leg and then subsequently becoming an amputee. When I ask people how they walk, they will generally describe putting one foot in front of the other, or perhaps shifting their weight from leg to leg as they propel their feet forward one at a time. If I probe them a bit more, they admit that they need to be able to see to walk, but that they generally look not down but ahead. Since I have no foot, lower leg, or knee on my right side, I cannot compensate well for what many others' bodies know (and which mine used to know), which is that when one steps on a pebble or a crack in the sidewalk, one *achieves* staying upright through a combination of touch (of the foot to the ground), sight, and vibrations (Sobchack, 2010). My ability to do this is limited, which means that I must walk while looking down and slightly ahead of my feet to anticipate every placement of my prosthetic foot. Hence I can walk or observe, but not both. Research on amputees using prosthetic limbs suggests that we come to incorporate prostheses into our sensorium due to "long-term exposure to discordant forms of sensory information; the visual, proprioceptive and tactile aspects of this prosthesis" (Murray, 2004, p. 964). Likewise, visually impaired people who regularly use canes incorporate them into an extension of their body image; the sensory complexity leads to new understandings of one's body. In addition to people with disabilities who use devices, athletes, dancers, and people whose work requires skilled use of their hands or learning to detect sounds or sights also come to have complex sensory experiences of their bodies.

Some of how people come to understand senses in ourselves and others is the result of learning to perceive and think about sensory inputs within a given culture. It is helpful to "differentiate *sensations* (that is, information routed via distributed nerves and sense-system clusters) from *sensuous dispositions* (the sociohistorical construction of the sensorium, its reproduction over time and its alteration through contexts and technologies)" (Paterson, 2009, p. 779).

Internationally renowned percussionist Evelyn Glennie, who is deaf, offers a useful way of understanding becoming attentive to vibrations, nonverbal cues, and the ways in which senses converge in everyday life (including music), when she explains how she is able to play percussion instruments as a deaf woman.

Hearing is basically a specialized form of touch. Sound is simply vibrating air which the ear picks up and converts to electrical signals, which are then interpreted by the brain. The sense of hearing is not the only sense that can do this, touch can do this too. If you are standing by the road and a large truck goes by, do you hear or feel the vibration? The answer is both.

(Glennie, 1993, ¶3)

Researchers can learn to appreciate the blurring and combining of senses in their field settings, in interviews, even in the ways in which we simultaneously think, feel our fingers press and release computer keys, see text appear on a screen, and hear what goes on around us as we write. Thus researchers could reflect, for instance, not only on how what participants see and what they taste work together to form the meanings of ceremonial foods served at a Jewish Seder meal to celebrate the Passover holiday, but also we could reflect "the other way around" to try to understand how participants divided up their experiences of eating Seder foods "into the categories of what is known by seeing and what is known by [tasting] because this would help me to understand how [a participant] gives culturally constructed meaning to these activities" (Pink, 2011, p. 267). The more researchers are aware of how our senses work in practice (beyond the five separate senses and corresponding organs model), the more possibilities manifest for inquiry and wonder.

Embodied Knowing

Researchers tend to act and speak as though knowledge is produced in the mind; that is, "the researcher's body is typically regarded as an instrument for sensory data collection, which is then rationalized and given meaning as knowledge by the mind" (Brady, 2011, p. 323). In resisting the mind–body split, researchers can understand embodiment not as actions and practices that our bodies do and that our minds subsequently make sense of, but rather as our whole body-selves making sense of the world and producing knowledge. Bodies are complex systems that *include* the brain and central nervous system but are not interpreted solely by them. Instead, knowing is a corporeal process that is tied up with our ontology, or way of "being-in-the-world" (Merleau-Ponty, 1962), such that "consciousness is always and only embodied, holistically integrated into the enfleshed subject," (Hoel, 2013, p. 35) and "[b]eing and knowing cannot be easily separated" (Longhurst, Ho, & Johnston, 2008, p. 208). At the same time, I resist a simple inversion of the hierarchy to privilege the body over the mind, thus reinforcing the dichotomy between emotional/imaginative and rational logics, merely reversing it (Shilling, 2012). We need "a way of re-thinking body-mind relations that complicates rather than erases demarcation between the two" (Barnacle, 2009, p. 28) and offers a way of making sense of knowing that includes the mind without making it the center of knowledge production and utilization.

FIGURE 1.2 Knowing through the Body.
Mixed media artwork by Joan BeJune and Laura Ellingson.

Courtesy of Joan BeJune (artist) and Gennette Lawrence (photographer).

A holistic understanding of embodied knowledge centers on being in the world through our bodies.

We encounter the world through our bodies and engage in some forms of preconceptual learning through our interactions with others using our senses (Barnacle, 2009).

> As animals we have bodies connected to the natural world, such that our consciousness and rationality are tied to our bodily orientations and interactions in and with our environment. Our embodiment is central to who we are, to what meaning is, and to our ability to draw rational inferences and to be creative.
>
> *(Johnson, 1987, p. xxxviii)*

And yet qualitative research does not typically acknowledge the ways in which all knowing is woven throughout our bodies. The silence of researchers' and participants' bodies is apparent within the constraints of traditional research report writing (Ellingson, 2006) and consequently omits vital information and insights: "When the body is erased in the process(ing) of scholarship, knowledge situated within the body is unavailable. Enfleshed knowledge is restricted by linguistic patterns of positivist dualism—mind/body, objective/subjective—that fix the body as an entity incapable of literacy" (Spry, 2001, p. 724). Thus, it is not that the mind makes (cognitive) knowledge out of the raw, perceptive (bodily) information. Instead, the body-mind system feels, thinks, experiences, and makes sense of sensory information holistically, which then may be rendered under a rational/cognitive guise via language. For example, Murray's (2004) study of prosthetic limb use describes how participants generated embodied knowledge through their daily practice of moving their bodies while incorporating into those movements different types of hand prostheses.

> Using a prosthetic becomes a form of knowing—an understanding which is achieved practically and corporeally. [Participant] at once describes the limits and potentiality of a prosthetic hand. While she is unable to perform complex motor acts with the prosthetic, relatively simple activities, such as holding a hymnbook, are made 'knowable' to her by virtue of the prosthesis. Knowledge becomes corporeal.
>
> *(Murray, 2004, p. 969)*

Researchers seek enfleshed knowledge (Spry, 2001) through *"learning in and as part of* the world, and seeking routes through which to *share or imaginatively empathize with* the actions of people in it" (Pink, 2011, p. 270; original emphasis). Often researchers have to try things ourselves to learn them, evoking the image of an ethnographer as an apprentice to participants whose embodied knowledge is of interest. "Ethnographers' bodies are not containers for the minds, but unfinished, open, rough-edged" (Hopwood, 2013, p. 239), and their open bodies are subjected to all sorts of practices and circumstances that alter what the ethnographer knows, not in her mind but in her whole body-self. Hopwood (2013) offers a great example of learning how to rock an infant to sleep during fieldwork at a family center, a caring practice his body-self had not known previously.

The notion that [participants] know with their minds and do with their bodies was wholly inadequate. . . . It was not just that by doing I came to understand. The doing was not just in my body, the understanding not just in my mind. The evolution of both was iterative, the locus of the knowledge and learning in mind or body indeterminate at any moment.

(Hopwood, 2013, pp. 237–238)

Hopwood's explanation exemplifies the degree to which knowing becomes part of our hands, arms, and chests that securely hold infants, part of our lips, tongue, mouth, and larynx that form sing-song noises of comfort. If asked how to hold a fretful child, we would likely adjust our arms and hands to show our answer, because this practice is known in gestures, posture, movement, and sounds—that is, in the body—rather than in language and rational thought.

Sticky Web of Culture

Following poststructuralist approaches to embodiment, culture is not merely the context of the body that shapes bodily perceptions of sensory input. Instead, culture is understood as *part of* the senses, with cultural norms and ideas, language, discourses, and practices all existing in a "sticky web" that cannot be separated from our embodied sense making processes (Rogers, 2003, p. 2). Thus the body itself both shapes and is shaped by society in complex ways, and one cannot access the body in a pure form, unaffected by culture (Shilling, 2012). The stickiness of the web offers a tactile image of researchers' and participants' bodies caught up in varying threads of cultural discourses that overlap and, twisting and turning, constrain and induce bodily movements and shapes, requiring significant force to break out of culturally determined norms and shaped by ongoing choices. In a study of physical (in)activity, the authors acknowledged the lenses through which they explained their athletic and everyday movements:

Even the most physical of sensations (like those caused by a dislocated hip, or a baseball bruise, or a long run in the hot sun) are always already discursive: they are felt through the experiences, hopes, desires, and languages available to the feeler.

(Avner, Bridel, Eales, Glenn, Walker, & Peers, 2014, p. 57)

At the same time, even though understood in and through discourse, bodies are also material. "The body may be surrounded by, perceived through, and at times self-managed on the basis of discourses, but it is *irreducible* to discourse" (Shilling, 2012, p. 85). Foucault de-emphasized the materialism of bodies and the phenomenology of the body in favor of a focus on its discursive properties. Ideally, qualitative researchers would frame bodies as both discursive *and* corporeal. Burns (2003), drawing on Probyn (1991), suggests that researchers conceptualize

> the body as simultaneously material/textual by adhering to notions of discourse as constitutive of regimes of truth about the body and as practices that shape the body and regulate embodied subjectivity. . . . [P]articular types of corporeality do not express a 'truth' about the body but rather 'articulate certain conditions of possibility' (Probyn, 1991: 114) among multiple possibilities. (p. 231)

In this way, embodiment is material/textual and continually in a state of *becoming*, never fixed or finalized (Deleuze & Guattari, 1987).

Discourse is more than just language, of course, but language is a key element of culture. Our bodies cannot be understood apart from our languaging of them. While bodies have material being, that materiality cannot be transparently understood. Instead, linguistic lenses filter all interpretations. Words do not spring forth from nowhere; we draw language from cultural reference points to construct categories, descriptions, and labels (Wittgenstein, 1953). Hence meanings of researchers' bodies and those of research participants are constructed in particular sociohistorical contexts that provide constraints of language resources. We make sense through our bodies and then reach for language to express ideas. "In the passage from the heard, seen, smelled, tasted, and touched to the told and the written, language has taken place" (Trinh, 1999, p. 263). Once language has taken place, meaning is created, assigned, even imposed on the body, and researchers' languaging of experience and ideas can be thought of neither as somehow reporting pure bodily experience nor as purely disembodied knowledge. Likewise, audiences always jointly construct meanings with researchers/authors when they read research reports in other, equally specific sociohistorical contexts.

All aspects of culture, including language, are inherently imbued with power: "power arises and is constituted within and through interpersonal relationships, social institutions, and in contexts where social interaction among various groups of people takes place" (Hoel, 2013, p. 35). So understanding power as embodied means focusing on how discourses of power position different types of bodies and how those bodies are read or made sense of by ourselves and others. In our efforts not to essentialize identities or be overly deterministic when discussing gender, sexuality, race, disability, and other social categories, theorists can mistakenly become so abstract in our understanding of the ways in which these categories come into being and are enacted through daily interactions, media, institutions, and organizations, that we can neglect to focus on the ways in which such intersecting categories have a material basis and corporeal consequences. When taking seriously how embodiment is part of any research, it becomes necessary not just to work the tension between discourse and materiality but also to consciously grapple with the ways in which power is manifested in both the discursive body and the material body and then implications of these power arrangements for real people. For example, in Scott's (2013) study of African American women navigating predominantly white institutions, she describes strategic choices black women make for their verbal and nonverbal communication to counter pervasive stereotypes about black women, including being uneducated, too loud, oversexed, angry, and having a lot of children. Her participants explained how they changed their vocabulary and syntax, tone of voice, posture, gestures, and clothing in order to resist the culturally dominant ideas about their identities at the intersection of race and gender (and class). They consciously navigated the sticky web of cultural stereotypes in order to succeed within predominantly white institutions, a frequently emotionally and physically exhausting process. Respondents also described overcompensating to try to distinguish themselves from the negative stereotypes, illustrating how difficult it can be to free oneself from the sticky threads that surround us in cultural webs of meaning around race, gender, and other intersectional identities.

Intersubjectivity

The concept of intersubjectivity helps to illuminate the common ground in which researchers and participants meet. This is not to say that researchers and participants end up with the same embodied experiences or a shared understanding of a given moment. However, our embodied experiences of one another are neither fully subjective (interior and generated autonomously), nor fully objective (exterior of all beings and providing a pure view of reality (what Haraway [1988] called the "God trick"), but as intermingled, reciprocal, and enmeshed. La Jevic and Springgay (2008), drawing on Merleau-Ponty (1968), describe intersubjectivity as involving a process in which

> each body/subject participates with other body/subjects, comingling and interpenetrating each other. Bodies bring other bodies into being without losing their own specificity, and each materializes itself without being contained. Rather than an understanding of self and other as oppositional, intersubjectivity becomes imbricated and reciprocal.
>
> *(p. 69)*

Such a relational, embodied understanding of self contrasts sharply with the independent, autonomous self prized in the United States and other individualistic societies. Imbricate means, roughly, to overlap, an idea that qualitative researchers can use to signify the ways in which participants' bodies depend upon one another for meaning in a given space.

Overlapping encompasses both sameness and difference: "the very idea of intersubjectivity presumes also that individuals differ and, therefore, that it always takes imagination to go beyond one's own experience" when trying to understand another's embodied experiences (Råsmark, Richt, & Rudebeck, 2014, ¶10).

> Our enfleshment throws into doubt the very sense of self and other as distinct entities and speaks to a folding over of flesh that creates the possibility of difference within a unified but undifferentiated medium (Shildrick 2013). In turning away from a knowing sovereign subject and insisting that we are all enveloped by the flesh of the world, Merleau-Ponty . . . [argues for] not a merging of subjectivities, but more a coming-together in difference, a matter of both convergence and divergence.
>
> *(Shildrick, 2015, p. 15)*

Intersubjectivity thus holds sameness and difference in productive tension, not resolving in favor of one or the other but embracing both.

Four aspects of significance for the understanding of intersubjectivity include lived space, body, time, and other (Van Manen, 1997). Each of these helps in understanding the intersubjectivity of how body-selves encounter each other. *Lived space* surrounds us and influences us as we act within spaces that shape how we feel, move, and understand our way of being. Thus the space in a hospital emergency department gives us a different set of emotional and sensory experiences of a sick child than would a child's bedroom. *Bodies* or embodied selves are highly interwoven with the presence and actions of others' bodies. Thus the intersubjective experiences of parents and their sick children differ significantly from the intersubjective experiences of sick children with emergency department nurses and physicians, for example. How bodies speak, touch, and otherwise interact with each other is essential to understanding meanings whether in a home, hospital, or other place. *Time* in intersubjectivity is not focused on clock time or linear passing of time, but instead on the lived experience of time, as going fast or slow, for example, or as painful or pleasurable. Time spent worrying about a sick child's uncertain prognosis may pass with excruciating slowness, for example. Finally, others are part of intersubjective experience because we can know ourselves only in relation to *others*—how they behave toward us, what they say (and do not say), how close and far away they remain. We come to construct a self and engage in ongoing revision of our embodied selves through interaction with others. Physicians come to know themselves as authoritative medical experts when they give orders that are promptly obeyed by nurses and other staff. These four aspects of intersubjectivity together can help researchers describe intersubjectivity between themselves and their participants and among participants (Shildrick, 2013; van Manen, 1997).

One autoethnographic study focused on weight loss experiences at the intersection of gender, race, and motherhood status to explore the ways in which cultural messages about obesity and health manifested within women's experiences individually and when compared and analyzed together (Johnson & Eaves, 2013). Their sense making clearly highlights intersubjective sense making. The researchers understood themselves as sharing a female identity but having different racial identities (one black, one white), and one was a mother and one not. Intersections were understood through both contrasting and shared embodied experiences of gendered and sexualized bodies and weight loss. Moreover, a study of breastfeeding women used the term "interembodied" because the intersubjective link was so strong between the women and their babies that the sense of the need to feed and the need to be fed became a deeply embodied experience of interdependence (Ryan, Todres, & Alexander, 2011).

In qualitative research, intersubjectivity emerges through imbricated sense making in ethnography and interview research and other forms of interactive data collection and analysis. Moreover, the concept of intersubjectivity points primarily not to linguistic or mental processes of memory making but to deeply embodied processes in which language and expressive (nonverbal) body language manifest through specific body-selves, in particular places and cultures. Researchers can attend critically to the embodied elements of their interactions and ongoing relationships with participants, attending to how their intersubjective relations not only influence each other but are mutually constitutive.

Actants (Nonhuman Objects)

When researchers do embodiment, we consider bodies and their contexts holistically, including objects they encounter. As previously established, bodies interact with, influence, and are affected by other bodies, and we know ourselves (and others) through this interactive process. The body is central to our capacity to exercise agency in the world, and we often use tools or objects when we act (Shilling, 2012); the "materiality of the field includes such things as human bodies, buildings, desks, books, spaces, policies, theories, practices, and other animate and inanimate objects. These materials are granted agential nature and undeniable affectivity, or an undeniable force in shaping inquiry" (Childers, 2013, p. 602). Actor-Network Theory (ANT) offers a useful perspective that goes beyond how human bodies interact with other human bodies to include animals, natural and made objects, and discourses, based on the philosophy of Latour (2005). According to ANT "management and experience of the body is *assembled* through its position in a complex *network* of material, technical, natural and ideational phenomena" (Shilling, 2012, p. 76; original emphasis). Other theorists have further developed ANT (a full explication is beyond the scope of this discussion; see Latour, 2005). ANT illuminates the ways in which bodies interact with, affect, and are affected by nonhuman objects, including inanimate objects (e.g., computers, knives, cars, foods) and the bodies of animals, birds, insects, plants, and other species. "Actants" may be understood as any nonhuman object or process that is part of a network of *intra-acting*, or co-constituting themselves in relation with humans and acts (Barad, 2007). An open and evolving process, "defined not by boundaries but openings . . . embodiment in the posthuman register is necessarily spliced with, and coconstituted by, networks of information, technology, and materiality" (St. Pierre, 2015, p. 341). That is, humans and objects do not exist unto themselves a priori and then interact to affect each other; rather, the humans and objects come to be through intra-action.

Posthumanist theory highlights the importance of actants and proposes "the radical decentering of the traditional sovereign, coherent and autonomous human in order to demonstrate how the human is always already evolving with, constituted by and constitutive of multiple forms of life and machines" (Nayar, 2014, p. 2). Objects also have forms of agency: the "taking up of the tools [of discourse] where they lie, where the very 'taking up' is enabled by the tool lying there" (Butler, 1990, p. 185). So understanding embodiment in research means paying attention to how all aspects of a context work, not just the humans, such that "[s]ubjectivity is the achievement of bodies and objects acting together" (Pink, 2009, pp. 51–55). This suggests a methodology in which researchers "analyze how bodies engage with their environments in ways that don't prioritize or privilege particular individual human senses or faculties over the vast array of nonhuman objects that shape and enable these faculties" (Ash & Gallacher, 2015, pp. 69–70).

An example of the power of actants is found in the rhetoric of gun rights activism in the U.S. and the well known contention that "Guns don't kill people; people kill people." This claim is both correct and wrong. Of course, human agency is integral to the possibility of people shooting people using guns, and even of designing and producing guns (and drones, which allow shooting by remote operators). However, guns and bullets also are actants; they have agency, or induce effects by materializing possibilities that are not possible for humans absent that object—a handgun, rifle, shotgun, or automatic weapon. Therefore, it is impossible to understand shootings without including within analysis the politics of the (lack of) regulation and use of the actants (weapons), including gendered patterns of gun violence and the federally mandated ban on researching gun violence (Sotirin, 2016; for a nuanced discussion of the crucial role of syringes as actants in illicit drug use, see Vitellone, 2010). My own study of communication during dialysis care demonstrates how dialysis machines as actants structure interactions among staff and patients, while cleansing patients' blood of certain waste materials (Ellingson, 2015). Manning's (2015a) study uses queer theory to highlight the complex intra-action of discourses of heteronormativity, girls' bodies, and purity rings. Furthermore, Latour (2003) contends "a strong distinction between humans and non-humans is no longer required for research purposes" (p. 78). That is, when making sense of how embodied actions and agencies in a network produce particular discourses, identities, or embodied states, all parts of a complex network must be considered, and human beings should not necessarily occupy the center of the analysis or be treated as inherently more important or as having more agency or more power than any actant simply because of their humanity.

Flux

Dominant cultural views of the body and self promote a relatively stable notion of identity and physicality. Yet bodies, selves, perceptions, language, and social structures are more productively understood as continually in process and movement. Martin and Kamberelis (2013), drawing upon Deleuze and Guattari (1987), suggest researchers embrace

> an ontology of becoming(s) rather than being. Reality is viewed as a continual process of flux or differentiation even though this fact is usually masked by powerful and pervasive illusory discourses of fixity, stability, and identity that have characterized most of western philosophy and theory since at least the Enlightenment. This ontology of becoming(s) enables (even urges) us to see things differently – in terms of what they might become rather than as they currently are.
>
> *(p. 670)*

Within this ontology of becoming, two types of effects persist. "Lines of articulation" maintain status quo arrangements by emphasizing hierarchy and normalizing discourses and practices, while "lines of flight" point to ways out of repression through decentering and dispersing discourses and practices. Lines of flight point to continual flux and possibility for change.

In addition to fluctuating in meaning on the discursive and institutional levels, bodies also undergo constant change on an individual level. People do not experience their bodies "in a way that is stable, straightforward, and permanently fixed . . . Instead, embodiment must be conceptualized as a highly complex, ever-shifting process, which is constituted not only by our social and cultural background, but also by the specific situations" in which we live as body-selves (El Refaie, 2014, pp. 110–111). The body should not be viewed as a naturalistic entity that is the origin of the True body; instead, bodies always already shape and are shaped by social norms, relationships, and institutions. "Bodies can be socially influenced, but they can also be seen to actively negotiate, adopt, or resist normalizing discourses. This is a process in flux . . . Lived bodies are . . . agential and productive, with a life of their own" (Harris, 2015, p. 9).

Likewise, brains are not fixed but exist in a continual state of flux known as *plasticity*; the brain is entangled with the surrounding world, adapting and changing as a result of our encounters (Schmitz & Höppner, 2014). This continual state of neurological flux is not well captured in neuroscience research because "brain images are snapshots of a certain moment of physical materiality" (Schmitz & Höppner, 2014, p. 5), and they cannot show processes, nor can they explain what led to brain development, nor the degree to which those developments are biologically and/or culturally determined. Yet these fixed moments of time are used routinely in neuroscience studies to study brains, as though brains themselves existed in static form. Brains, culture, and objects all intra-act to co-constitute each other (Barad, 2007). Feminist neuroscientists and other scholars of embodiment push back at this "neuroreductionism" with more nuanced and fluid conceptualizations of embodiment (Pitts-Taylor, 2015, p. 22).

How aware people are of our own embodied selves and embodied ways of being in the world is not a simple question. For example, a study of people living with significant hand disabilities emphasized the process by which shifting states of embodiment exist over time and in different contexts.

> [S]ome participants' states of embodiment were not well-described as either the object body or as cultivated immediacy but as something in between these two states . . . Attaining and maintaining a new state of body-self unity often also involved . . . the modification of expectations, use of humor, downward comparisons, trying to ignore or forget the hand condition, and focusing on abilities and on positive rather than negative things.
> *(Hudak, McKeever, & Wright, 2007, p. 39)*

Moreover, the sense of embodiment for people with prosthetic hands was deeply influenced by the ways in which they were treated by others around them: "the views of others—whether real or perceived—seem to make it difficult for an individual to remain unself-conscious of his or her body and to sustain an undisrupted sense of self" (Hudak, McKeever, & Wright, 2007, p. 40).

Just as no static bodies exist, no permanent or unchanging fixes can be made to bodies. Human bodies are both fragile and resilient, and any fix, adjustment, or work-around is workable for a moment, not forever. In the same way that most people take sight for granted, we also tend to take for granted that we have fixed a problem. When we are sick and we go to the doctor, we expect medical care to fix or at least ameliorate our troubles. Yet those fixes will not last – fillings and crowns on teeth break, surgically repaired knees develop arthritis, and lower back pain tends to recur following treatment. Change comes not just to our bodies as they are, but in response to any solutions we find to what ails us. We need to find ways to conceptualize bodies as contingent, as always already in the process of becoming different than they are. A study of disability and femininity among women who *almost* pass as able-bodied demonstrates that contingency is inherent to the daily doing of (disabled) female bodies; "gender is always in a state of performance—pursued, longed for, but forever unfinished because perfected gender performance eludes us through its only imagined existence" (Scott, 2015, p. 238). Gender and disability intersect to make possible certain performances of self but no final, perfected performance of femininity (nor of "super-crip," see Murphy, 2001).

Moreover, part of the fluctuation of bodies is a multiplicity and indeterminacy of a singular "*the* body"—even though I, too, resort to that shorthand phrase when discussing the admittedly dynamic state of embodiment. Instead bod*ies* must be taken seriously as multiple, diverse, situated, and becoming. The myth of a singular body privileges those bodies upon whom the status quo confers the most privilege—white, male, heterosexual, affluent, First World citizen, able-bodied—by generalizing the experiences of the elite as stable, normative, and as the ideal to which fixed notions of nonwhite, female, LGBTQ, poor, Third World, and disabled bodies will inevitably be found to fail to conform (Minow, 1990). Understanding that all bodies are always in the process of becoming other than they are opens possibilities for bodies to be valued for their future manifestations rather than degraded on the basis of fixed stereotypes based on stale social categories.

Practices for Embodying Research

In this chapter, I sketched a model of embodiment that draws on a wide range of theoretical traditions that overlap and have some common threads, but which also have important differences in their traditions and trajectories. Playing with each of the concepts can be helpful as you begin the process of becoming more deliberate in your engagement of embodiment throughout your research processes.

Learn your Sensorium

Pink (2015) articulates and advocates sensory ethnography as a method. As preparation for sensory fieldwork, Pink suggests that ethnographers learn their own sensoriums through careful reflection on their participation in sensory cultures. Try attending to and describing a single interaction or a specific task from your everyday life in an integrated, holistic sensory way. Recall the discussion of how what we see affects what we taste, what we hear affects what we see, and so on; our senses are not discrete but integrated in our experiences. Try to understand your sensorium in its integrated, generative creativity. Draw on figurative language to express your sensory perceptions, experimenting with simile (e.g., the blood flowed through the dialysis machine like a red river under a strong current), metaphor (e.g., the vile odor was a punch to the head), or hyperbole (e.g., the dialysis technician was the tallest man in the world as he towered over the tiny, elderly patient). Be poetic and imaginative as you try to taste colors and hear textures, and so on. Think critically about what you have learned to sense and about what your socialization may prevent you from recognizing as significant. Consider what you have been socialized to think of as disgusting, beautiful, offensive, and pitiful and how these judgments relate to sensory perception of smells, sights, sounds, tastes, contexts, and so on.

Chart Your Body

Using tools made possible by the quantified self movement (QS), experiment with gaining self-knowledge through self-tracking—i.e., wearable monitors and smart phone apps that track a person's bodily activities and movements, heart rate, and other biometric information (e.g., Nafus & Sherman, 2014). While mindful of (and concurring with) critiques of QS practices (e.g., Lupton, 2015), I nonetheless encourage you to use inexpensive or free apps (or ones you already own) to chart various body functions for a week or two—weight, ovulation cycle, heart rate, exercise, sleep patterns, or whatever interests you—so that you build a set of quantitative data about yourself. At the same time or during the following week, construct a qualitative record—in a journal or table format—in which you construct a qualitative record about aspects of your body-self. Make brief descriptions throughout the day of energy levels, muscle tension, hunger sensations, pain, perception of stress in your body, and so on. Then consider what you know and how you feel about your body in response the quantitative record, comparing that to the knowledge and feelings sparked by the qualitative record. Reflect on how self-tracking shapes how you know your body and how your knowledge of your body shapes how and what you track.

Possibilities of Blended Senses

Consider the discussion of vibration, touch, and hearing in the sensorium section of this chapter, then listen to Rawlins' (2013) performance text, "Sample," available at the online performance journal, *Liminalities*: http://liminalities.net/9–3/sample.html. In this piece, Rawlins performs a song he wrote that includes, among other ideas and language, the abstract of a research article published in a gerontology journal. The wonderful discordance in hearing Rawlins sing the research jargon along with music played on a guitar is provocative and inspiring; it opens up possibilities for understanding strengths and limitations of both genres—music and research reports—and prompts us to ask ourselves as researchers why we do not typically sing our data or findings. Consider how different senses and modes enable you to experience your data or findings and then experiment. Following Rawlins' lead, try to dance your findings, cook them as a feast, or make a scrapbook of them, then reflect on your sensory experiences during these different ways of encountering data.

Tone

Consider the tone of your research setting, whether it is the room in which you conduct interviews, your ethnographic field site(s), or the office in which you construct and analyze data. "[T]one is a way of thinking about how vibrations are organized with particular sensory effects in mind. . . . the tone of an environment or object primes specific responses or actions to potentially occur" (Ash & Gallacher, 2015, p. 79). Attending to tone can help attune participant observers to how tone can "shape the kinds of possible futures that can emerge from a situation, albeit in an open and nondeterministic way" (Ash & Gallacher,

2015, p. 80). Ethnographers can pay attention to the myriad elements of settings far more than we typically would to see how vibrations and tone circulate through the human and nonhuman objects. When you talk to people informally or in interviews, ask questions about how people perceive surrounding objects and how their senses are engaged in that setting.

Consider Animal Bodies

Need an excuse to watch more cat videos on the Internet? I highly recommend that qualitative researchers learn more about doing bodies (as an alternative to having bodies) by watching a cat or dog play, eat, sleep, or interact with others (go on YouTube if you don't have a pet at home; also, many animal shelters and rescue centers are grateful for volunteers to play with or walk animals). Animals always do their bodies; they are present with their movements and with their stillness. Cats may coyly ignore a toy mouse and then pounce, hunt, stalk, or creep up on it. Practice describing a cat's or other animal's movements, expressions, and sounds, either by watching a video or by interacting with a pet. During the writing of this book, one of my cats, who has long, fluffy fur, sat on my leg, on a table next to my computer, across my chest and shoulder, and next to me on a couch, providing both inspiration for reflecting on embodiment and tactile, visual, and auditory stimuli. For example, I practiced describing all of the tactile sensations I could identify as I touched him with a hand or felt him elsewhere on my body: warmth, softness, a sensation of weight and pressure from his body, the smooth texture of his fur, differing vibrations from his purring, snoring, and breathing. Such practice is both fun and helpful in learning to enrich your perceptive vocabulary and capacity for differentiating between subtle cues given by your participants' bodies.

Visual Journaling

"A/r/tography," which involves reflexive, visual journaling, is another rich possibility for a research process that focuses on illuminating the embodiment of researchers and participants.

> A/r/tography as visual journaling emphasizes the unthought, the spaces that are messy, uncomfortable, and complicated. Through the intersection of image and text, students are asked to create meaning from disparate and fragmented sources, to search for hidden experiences that are not apparent on the surface, and . . . [to] assemble the frameworks of understandings.
> *(La Jevic & Springgay, 2008, pp. 84–85)*

A/r/tography embraces messy spaces where text and images come together (or refuse to connect) and opens up a great way of making sense through embodied sensations and perceptions that are not captured through text alone. The material process of assembling text fragments and images—whether done digitally on a computer screen or using paper, scissors, and glue—stimulates kinesthetic and visual modes of knowing that can complement and contrast with explicit, implicit, and ironic meanings of text fragments as they intra-act (Barad, 2007) to form a visual journal entry. Experiment with constructing a visual journal by typing up reflections in response to questions, prompts, or a particular research interactions, and then digitally or with scissors cut up the reflection and recombine it with images pulled off the Internet, or out of magazines, flyers, books, or other printed materials. This can be done low-budget and low-tech, or if you enjoy the process and find it meaningful, consider investing in a mixed media journal.

Everyday Life and the Body

What do we come to take for granted as part of everyday life in our dominant cultures? What is unquestioned and assumed in day to day activities?

> Daily life eludes [researchers], even and especially when made the object of attention, because the very registers of science are set to tune it out. The everyday is in this sense antithetical to analysis; it consists of just that which remains once analysis has run its course.
>
> *(Hall, Lashua, & Coffey, 2008, p. 1023)*

Doing embodiment in data collection means learning to stop tuning out the everyday and attend to it actively. Describe participants' routine behaviors, movements, and practices, reaching for details of what you sense is typical in their everyday world. As an exercise, try to describe in minute detail how a participant's body-self carries out a specific, common task to bring to light micro-actions. For example, in a dialysis clinic, I noticed the routine practice of one staff member calling out to another who was elsewhere in the room, "Thanks!" for no reason that I could perceive initially (Ellingson, 2011b). In time, I noted the practice of technicians stopping to press reset buttons on the computer screens of dialysis machines as they passed by them, even though the machine was not one of those to which they were assigned that shift. When the assigned technician noticed this quick assistance from elsewhere in the room (which saved them having to move to the machine and press the buttons themselves), they called out a "thanks" to their helpful colleagues. This everyday form of communication expressed appreciation, fostered collaboration, and saved time and energy for technicians who did not have to interrupt other tasks to reset machines. Such an everyday practice was not mentioned to me by staff in interviews, but when I described what I had observed, technicians affirmed that this was a practice they did largely without thinking about it.

Own What Your Body Does

Accounting for your embodied research processes is vital to constructing a comprehensive methods section in an article or an appendix or flyer to accompany arts-based research. Moreover, such a roadmap assists others who may seek to follow your lead. One of my persistent criticisms of qualitative researchers is their unwillingness or inability to include the mistakes and misdirections that inevitably constitute actual research; we neaten up our stories of "what happened" to sound credible and get published (see Ellingson, 2009a, pp. 119–120). The rejection of the messy details of research is a legacy of positivism. Explains one researcher, "Let's repress the mess: that is the policy. Let's Other it" (Law, 2007, p. 602). To construct findings, researchers have to deny all the messy stuff that does not fit. Law calls this "common-sense realism" that guides our construction of findings, an Othering of "a reality that was multiple, slippery and fuzzy. Indefinite" (Law, 2007, p. 603). I do not believe that most qualitative researchers deliberately lie, but we often omit our embodied messes in our research sites. Yet researchers need models of how to negotiate uncomfortable circumstances. Moreover, the mystery of how we moved from daily life to autoethnographic narrative, or from coding to themes, should also be explained. Space limits in journals remain, as does pressure to neaten up accounts. However, researchers can write full accounts as part of their own research processes without including them in publications, sneak a subversively embodied, messy sentence or two into a methods section or endnote, and consider writing a separate "confessional tale" (Van Maanen, 2011).

2

DESIGNING BODIES
Planning for Conscious Embodiment

Doing Legwork: Speaking Ethics

Sighing, I turned around and limped out of the waiting room, through the hallway lined on both sides with examination rooms, and back into the staff-only desk area, where I would leave my purse and prepare my recording equipment. I tried to scratch at an itch on my hip through my skirt, underwear, and stockings but didn't succeed at alleviating the discomfort (I eventually learned it was best simply not to wear stockings). Living in the South seemed to involve an endless cycle of boiling alive in the hot sun and humidity, followed by shivering with clammy skin as the sweat dried on my skin, only to endure the process again when I moved from one building to another. Over the course of a day spent alternately sweating and shivering, my skin got very itchy. I was definitely a native of my beloved northeastern New England, suffering from my choice of a PhD program in the deep South's hot, humid, bug infested, alligator ridden, mildewed territory, albeit with some great Southern food and some wonderful people. I took a deep breath and walked forward some more, winding my way through the crowded, narrow passageway.

I had failed again on Tuesday to get any of the morning's patients to agree to allow me to audiotape them as part of my qualitative study of an interdisciplinary, geriatric oncology team that provided a comprehensive review and treatment plan for patients over the age of 70 diagnosed with cancer. I felt my shoulders stiffen and my voice shake every time I tried to convince a patient that I was a competent, professional researcher, rather than a nervous graduate student.

Today was another new day, and I entered the waiting room again, zeroing in on a likely patient from my list. Approaching her, I smiled and said with what I hoped was calm professionalism, "Hi! My name is Laura Ellingson. Are you Mrs. Robellin?"

"Ah, yes," replied the woman cautiously.

"Hi! My name is Laura Ellingson," I began, smiling widely. "I am a PhD candidate and a researcher working with Dr. Armani. I'm here to see if you would be willing to enroll in a study we are doing on communication throughout the process of the comprehensive geriatric assessment of patients. If you are interested, I would follow you throughout your visit here in the clinic. You would still have the same health care providers, for the same time period. I would come with you and record your discussions. Would you be interested in discussing the consent form for this study?"

I held my back stiff and straight, trying to schooling my face to appear calm and professional as she stammered, "N-no, thank you."

"Okay, thanks. Have a good day," I said, trying not to show my disappointment.

I smiled at yet another senior adult, this one a man whose gray hair stood in military precision, his back straight, his nervousness betrayed by one bouncing knee. Feeling like a fraud, I tried again to sound confident, yet kind, authoritative yet approachable. Strike two.

The next patient on my list was a woman who about the same age as my own grandmother, with the same white, fluffy hairstyle. She smiled as I approached, and I felt my body relax, my shoulders resuming their normal position. I smiled warmly and in a deliberately softer, higher pitched voice than my "professional" tone, I began. "Hi! My name is Laura. And I'm doing a project for school here at Dr. Armani's clinic." After confirming her identity, I continued. "I'm going to grad school in communication, and I'm interested in how the staff and patients talk with each other. If you might be interested in helping me, you wouldn't have to do anything differently then you were going to do anyway. I would just come with you and record you talking with each of the staff members, including Dr. Armani."

The couple both smiled warmly, and Mrs. Dennis said, "Isn't that wonderful, Bob? She's going to school."

"It sure is!" agreed Mr. Dennis, smiling and holding his hand out to shake mine.

I beamed at both of them, keeping my voice soft and friendly. "That's right, thanks. If you might want to help me out, we could go over this consent form together, so that my professor would know that I followed all the proper guidelines. And then you could decide if you would like to participate or not. And of course, it's completely fine if you decide that you would rather not do it. Either way is totally fine," I said, nodding my head for emphasis. I scanned both people's faces for signs of concern or confusion, but saw only friendly smiles as they both looked directly at me.

"Well, of course we will help you," said Mrs. Dennis, nodding. "I think it's just terrific that so many girls get to go to college today. Back in my day, we mostly didn't get to, you know."

"That's wonderful! Thanks so much," I said, relief coursing through my body. "Let me just show you this paperwork, and I can answer any questions you might have."

I spent a pleasant morning with Mr. and Mrs. Dennis, who chatted with me cheerfully in the lulls between interactions with the geriatric team members.

Questioning the Body

Janesick (2000) reminded scholars that all qualitative research begins with a question based on a curiosity or passion to know more about something. Research questions for studies can take embodiment seriously as their topic focus, as part of their reflexive research processes, or as a significant goal for representation of findings both within academia and in communities outside of higher education. In doing so, these research questions can address goals of social justice and shed light on the unspoken, taken for granted aspects of life, becoming "[q]uestions that afflict the comfortable and comfort the afflicted" (Janesick, 2000, p. 382; see also Preissle, 2007 for a discussion of feminist research questions). I tell my students that qualitative research questions begin with *how* or *what* but rarely *why*, especially when we inquire into the body. Why implies causality or directionality of forces, which is a question that typically can be answered only after more descriptive analysis focuses on the who, what, when, and where of a group under examination. The unique strength of qualitative research is the capacity to richly explore meanings, moments, and becomings. We ask: How does this practice work in and through bodies? What is going on in this place? What does this embodied identity mean to whom? In what ways is this discourse manifesting through these practices? What happens when this combination of bodies, place, objects, and discourses intra-act at this time? These are questions that focus on the body.

A study of post-bariatric surgery patients' understanding of their embodied eating practices, for example, asked two research questions: "How do patients experience eating and a change of eating practices in the long term after bariatric surgery? How do they describe the body in relation to eating?" (Natvik, Gjengedal, Moltu, & Råheim, 2014, 1701). Researchers interviewed 14 patients at least five years post surgery to learn about their eating and exercise practices, their experiences of their bodies, their evolving identities, and others' embodied responses to the participants' bodies. A study of enactment of mind–body dualisms in Wall Street banks—that is, employees' denial of their physical needs, experience of cognitive work, and embodied consequences of chronic stress—in a 13-year-long ethnography asked the following:

1. How is Cartesian dualism instantiated in modern knowledge-based organizations, including in practices, narratives, and forms of embodiment?
2. How do these situated elements (practices, narratives, forms of embodiment) coevolve over time and across different contexts?
3. What are the evolving consequences for participants?

(Michel, 2015, p. S45)

These questions delve into mind–body connections of workers in a particular environment that makes heavy demands on workers' bodies while simultaneously ignoring those bodies as mere vehicles for the highly complex, time-sensitive cognitive labor of their brains.

Research questions often shift and evolve with, rather than precede, fieldwork and interviews. Janesick (2000) uses the embodied metaphor of dance to explain the fluid process of qualitative design, drawing on metaphors of the minuet as a dance with precise, predetermined, sequenced steps, and of improvisational dance as a form that requires preparation, exploration, and illumination in response to particular moments in music. All that spontaneity requires lots of learning, training, and preparation, in order to be ready for the fantastical moments of performance. Likewise, consideration of embodiment is critical to preparation for designing a qualitative study. "Like the dancer and the choreographer, the qualitative researcher must be in tune with the body: The eyes must be taught to see, the ears must be taught to hear, and so on" (Janesick, 2000, p. 386). Researchers should assume that their plans will be fluid, the process nonlinear and iterative, and opportunities and relationships evolving (Tracy, 2013). All research questions have elements of embodiment implicit, and some have explicit attention to bodies; at the same time, researchers' and participants' bodies are always integral to any research question posed, so embodiment is always part of any critical or qualitative study, even if not addressed explicitly.

Designing Embodied Qualitative Research: A Continuum Approach

When considering how awareness of and attention to embodiment will be part of a research project, I find it helpful to envision the field of qualitative, interpretive, and critical methods as existing along a continuum, rather than thinking in terms of dichotomies between science and art, hard and soft, mind and body, rigor and creativity, etc. Dichotomies limit our imaginations. Dichotomous thinking remains pervasive within methodological debates (Gergen, 1994), as evidenced strongly in the quantitative/qualitative divide, and even within the qualitative field (Atkinson, 2006; Ellis & Bochner, 2006). Moreover, as explained in the introduction to this book, dichotomies of the body are part and parcel of these entrenched patterns: rational mind/emotional body, male mind/female body, moral mind/indulgent (even sinful) body, objective mind/subjective body, knowing mind/sensing body, higher mind/lower body (Ellingson & Ellis, 2008).

Many qualitative researchers, if not most, will affirm that art and science are ends of a continuum, not a dichotomy; yet in practice, the use of the methodological "other" to legitimate our particular methodological and paradigmatic preferences leads to a reification of dichotomous ways of conceptualizing and writing about qualitative research practices (Ellingson, 2011a). Attention to embodiment can help researchers to more fully grasp the wide and nuanced range between art and science by grounding difference not only in theoretical and methodological concerns but in the material world and the ways in which different forms of research draw from and influence the bodies of researchers, participants, and audiences. Researchers typically understand that the minds and thoughts of researchers and participants are involved, but the notion that researchers' and participants' body-selves are both directly and indirectly implicated in all stages of research adds an ontological sensibility; that is, it sheds light on ways of being in the (corporeal) world as inherently part of what it means to create knowledge.

Intersecting with the art/science continuum are critical theory "lines of flight" (Deleuze & Guattari, 1987) which can be used to frame an analysis or to enrich the theoretical frame grounded in another mode. Mapping the body requires keen attention to where researchers situate themselves, what their assumptions are about research practices, participants, and truths, all of which affect and are affected by researchers' choice to pay explicit attention to issues of embodiment. Cutting across the continuum are the critical, theoretical, and ideological critiques of methods based on poststructuralist theory, which are widely discussed in qualitative methods journals, some of which call for a radical reconceptualization of data, analysis, and representation, especially as they relate to the body, (e.g., Mazzei & McCoy, 2010; St. Pierre, & Jackson, 2014). This growing and intriguing area of the qualitative field has inspired extensive scholarly conversation and reflexive consideration, yet it has not sparked a sea change in the dominant approaches to qualitative methods in published (particularly grant funded) research (Koro-Ljungberg & MacLure, 2013). Poststructuralist theory has had a tremendous influence on my own understanding of what I do when my body-self does research with other embodied selves. However, I do not see the critical insights, useful concepts, and productive contributions to research that this and related bodies of theory offer as necessitating a radical shift in my (admittedly eclectic) research practices or the abandonment of any particular form of (modernist/postpositivist) analysis or representation. Rather, I understand these contributions as reframing possibilities, sparking lines of flight that can be followed to expand opportunities for sense making.

Building upon Ellis's (2004) representation of the two ends of the qualitative continuum (i.e., art and science) and the analytic mapping of the continuum developed in Ellis and Ellingson (2000), I envisioned the continuum as having three main areas, with infinite possibilities for blending and moving among them (Ellingson, 2009a). The goals, questions posed, methods, writing styles, vocabularies, roles of researchers, and criteria for evaluation vary across the continuum as one moves from a realist/positivist social science stance on one end, through a social constructionist/interpretive middle ground, to an artistic/performative/ postmodern paradigm on the other end. No firm boundaries delineate the precise borders of these regions of the continuum. Furthermore, terms of demarcation and description used throughout the continuum (e.g., interpretive, postpositivist) continue to be suspect and contestable; use of terminology in qualitative methods remains notoriously inconsistent across disciplines, paradigms, and methodological communities, with new terms arising continually (e.g., Gubrium & Holstein, 1997; Potter, 1996; Denzin & Lincoln, 2011).

I now sketch the science, interpretive, and artistic areas of the continuum, along with a brief review of both traditional and more contemporary approaches to spanning the boundaries among those areas. I discuss the ways in which embodiment is constituted in and through research practices and how the body is (un)acknowledged in different paradigms of knowledge construction across the continuum.

Science

Anchoring one end of the qualitative continuum, researchers emphasize valid and reliable knowledge generated by neutral researchers utilizing scientific methods to discover universal Truth, reflecting a positivist epistemology (Ellingson, 2011a). Under positivism, the body is reduced to a mechanistic function of the mind. Historically, social scientists understood positivism as reflected in a "realist ontology, objective epistemology, and value-free axiology" (Miller, 2000, p. 57). Few qualitative researchers currently embrace an absolute faith in positivism, however. Many postpositivists, or researchers who believe that achievement of objectivity and value-free inquiry are not possible, nonetheless embrace the goal of producing generalizable knowledge through realist methods and minimization of researcher bias, with objectivity as a "regulatory ideal" rather than an attainable goal (Guba & Lincoln, 1994). Research in this tradition utilizes precise, prescribed processes and produces social scientific reports that enable researchers to make generalizable claims about the social phenomenon within particular populations under examination (Tracy, 2013).

In this region of the continuum, embodiment of researchers and participants is not part of the overt processes documented in research reports. Traditional positivist and many postpositivist methods in social sciences have historically placed great emphasis on "abstract cognitive inquiry" and considered corporeality to be a potential contaminant of data. Note that this approach to inquiry was also constituted by the "founding fathers" of social sciences who were predominantly privileged men for whom the risks of embodiment – slavery and racist violence for people of color, childbirth and sexual assault for women, hunger and cold for people in poverty – were not first on their minds (Shilling, 2012).

Thus embodiment was reduced to an issue of reporting participants' demographic categories, which are framed as unproblematic categories. That is, participants' sex or gender, racial identification, age, sexual orientation/expression, and other categories relevant to the research topic are provided by participants and reported in aggregated form. For example, an intriguing qualitative study of resilience and identity among disaster-relief workers, published in a traditional academic journal, described participants using basic demographic categories.

Interview participants included 4 full-time and 1 half-time staff member and 18 volunteers from the not-for-profit's chapter who self-selected themselves into this study by responding to the first author's emails or requests made during a monthly meeting (N=23, males=11, females=12). Because two of the four staff members were volunteers when they started, we use the term "disaster-relief workers," "members," and "volunteers" interchangeably to describe our participants. Demographic data collected through an online survey indicated that participants averaged 51–61 years of age (on a response set using decades ranging from 20 to 70 years and older), were primarily unpaid volunteers, Caucasian, had master's degrees, and had been with the chapter for 2–5 years.

(Agarwal & Buzzanell, 2015, pp. 412–413)

The researchers provided embodied differences as background information only, as context for their analysis of resilience for workers in a nonprofit organization and in relationships with the people that they serve during disasters. This example reflects a useful but very limited way of framing bodies. Two further options exist for those who choose to write in the traditional report format that is rooted in (post)positivist assumptions. First, researchers can conduct multiple analyses of a data set representing different regions of the methodological continuum, incorporating various forms of critical theory, and/or incorporating aesthetic or narrative sensibilities that illuminate bodies, embodied practices, and the researcher's body-self (Ellingson, 2009a; Harter, 2013a), even though they also have published results framed in a traditional science format that offers only implicit traces of the body. Creative secondary or subsequent analyses can address embodiment explicitly in a variety of ways, as detailed below in the discussions of the middle and art regions of the continuum. Second, researchers can benefit significantly from incorporating awareness of embodiment of research practices in a research log or reflexive journal, even if the topics explored are used only to enrich the researcher's thinking and are never explicitly shared in publications or other public representations. While sense making and writing are intertwined, researchers can discipline themselves to write according to strict postpositivist writing conventions either because they find them useful or because they wish to situate their work within an ongoing scholarly conversation that is taking place in spaces that generally are not open to other ways of reporting findings, called guerilla scholarship tactics (Ellingson, 2009a; Rawlins, 2007), e.g., for innovative theorizing on embodiment in qualitative management research, see Stewart, Gapp, & Harwood, 2017).

Art

At the other end of the continuum, artful researchers value humanistic, openly subjective knowledge, such as that which is embodied in stories, poetry, photography, performance, and painting (Ellis, 2004). Researchers may study their own lives and/or those of intimate others, community members, or strangers. Among artistic qualitative researchers, truths are multiple, fluctuating, ambiguous, and grounded in lived bodies. Autoethnographers, performance artists, and arts-based research practitioners embrace aesthetics, evocation of emotion, and negotiation of identity as equally or more important than illumination of a particular topic (Richardson, 2000). Literary standards of truthfulness in storytelling (i.e., verisimilitude) replace those of social scientific truth (Ellis, 2004). Like all art, creative social science representations enable us to learn about ourselves, each other, and the world through encountering the unique lens of a passionate rendering of reality in a moving, aesthetic expression. In artful modes, bodies figure quite centrally and explicitly in all stages of research, including and even especially the bodies of researcher-artists and the intersection of their bodies with participants' bodies in specific spaces and places. At the same time, artful researchers also confront limitations. The openness and multiplicity of artful representations of bodies, especially bodies interacting with one another, are not inherently novel or liberating, and may even inadvertently reinforce stereotypes or lack sufficient attention to embodied intersectionality, just as any form of research might.

Analysis and representation merge to a great degree as art is produced on the basis of experiences, data, and other empirical materials. A tremendous variety of artistic practices that engage embodiment exists, with new forms continually arising. *Autoethnography* is research, writing, story, and method that connect the autobiographical to the cultural, social, and political through the study of a culture or phenomenon of which one is a part, integrated with relational and personal experiences (Ellingson & Ellis, 2008; Ellis & Rawicki, 2013; Holman Jones, Adams, & Ellis, 2016). In autoethnography, the researcher's "reading of embodied sociotextual pelts becomes the written semantic interpretation of [her] own somatic experience" (Spry, 2001, p. 721). Autoethnographers typically produce emotionally evocative accounts that highlight the embodied experience of emotion that encompasses mind, body, and spirit. For example, Boylorn (2011), an African American woman, shares stories and poems about race, identity, and racism in teaching and in public interactions with strangers. She reflects on the ways in which her embodied self is racialized by others: "As my racial identity is inextricable from my researcher self, focusing on stories of difference and race helps me reflect and question the nuances of difference in everyday life" (Boylorn, 2011, p. 184).

Narratives constructed from fieldnotes, interview transcripts, personal experiences, or other materials enable readers to think and feel *with* a story, rather than explicitly analyzing its meaning (Frank, 1995). Parry (2006) constructed short stories based upon her interviews with women negotiating pregnancy, birth, and midwifery (for further narrative exemplars, see: Berry, 2016; Bochner, 2014; Defenbaugh, 2011, 2013; Holman Jones, 2016; Lindemann, 2012; O'Connell, 2016). The creative analytic practice of *poetic representation of* findings provides rich modes for artistic expression of research (Faulkner, 2009, 2014; Glesne, 1997; Guzzardo, Adams, Todorova, & Falcón, 2016; Hopkins, 2015; Richardson, 1992). Many researchers create live *performances* based on autoethnography, ethnographic fieldnotes, and/or interviews that engage audiences and invite them to connect to and empathize with others (which may then be recorded) (Rawlins, 2013; Spry, 2001; and check out the online performance journal *Liminalties* and the online publication space of *Women & Language*). Many *video* representations of empirical materials involve partici-patory methods, ideally empowering participants to act on their own behalf. One research team developed a dramatic production exploring personhood in

FIGURE 2.1 Eggstraordinary Story Art from the Scraps of the Heart Project's Creative HeARTs Workshop.
University of Denver Communication Professor Erin Willer leads creative workshops for parents who have lost a baby.

Courtesy of Erin Willer (photographer).

patients with Alzheimer's disease that is used as a teaching tool with medical students (Kontos & Naglie, 2007; see also Gray & Sinding, 2002). Another form of participatory method engages participants as artists, expressing themselves through collaborative artwork. The example in Figure 2.1 shows participatory artwork generated in a workshop led by Erin Willer, whose research engages parents and families who have experienced the loss of a baby.

Throughout all of these, details of artful bodies are represented as central to experience and sense making. With artistic standards for evaluation, bodies can be (re)interpreted as sites of knowledge production (Spry, 2001; Hesse-Biber, 2007).

Interpretive

The middle ground of qualitative inquiry remains my preferred (play)ground as a feminist, qualitative methodologist. In this region sits not merely work that is "not art" or "not science," but work that features description, exposition, analysis, aesthetics, insight, theory, and critique, blending elements of art and science or transcending those categories entirely (Ellingson, 2009a). Interpretive work may weave attention to embodiment in intriguing ways that make the best of both the structured analysis and conventional written format. Interpretive scholarship often (but certainly not always) incorporates attention to researchers' standpoint, power and politics surrounding participants and researchers' relationships with them, and more flexible expectations for writing to illuminate the material elements of analysis.

First person voice generally used in middle ground qualitative research departs from the passive voice that characterizes positivist work, emphasizing intersubjectivity of researchers and participants (and their social worlds), but without going all the way to embrace the avowed subjectivity of art. Interpretive researchers acknowledge claims of truth as contingent because

> meaning is always in flux and continuously changing as people interact and exchange ideas. Because of this, interpretive qualitative researchers focus on how people interact in, with, and to create a social scene. That differs from postpositive studies that often explore inherent meanings that might be found in a person, object, or idea; or that focus on a person's interpretation of people, objects, or ideas. Rather, interpretive qualitative research focuses on meaning making as a reflexive, complex, and continuous process.
>
> *(Manning & Kunkel, 2014, p. 3)*

Rather than ignoring the ways in which researchers actively construct findings or seeking to minimize bias, middle ground qualitative researchers often reflect upon the intersections of their embodied standpoints with those of participants and others to shed light on how race, class, gender, dis/ability, sexuality, and other embodied identities and experiences interact with, and shape research processes and results. For example, Edley (2000) wrote, "[I]t is important to make explicit my own standpoint. I am a 30-something feminist, White, middle-class, Protestant, heterosexual woman who is married and has a young son" (p. 300n2). Likewise, in our qualitative study of communication among aunts, nieces, and nephews within extended family and chosen family networks, my collaborator and I acknowledged our standpoints as contributing to our interpretive analysis.

> We are both European American women who are, or have been, in committed heterosexual partnerships. Patty has four children, while Laura has chosen not to be a mother. We are both nieces and aunts; our experiences of being aunted, and of aunting our nieces and nephews, share some similarities and yet diverge in many ways. Being an aunt is a crucial nurturing role in Laura's life, while it takes a distant second place to mothering and sistering in Patty's life. Laura's aunt Joan is one of her closest friends, and she has enjoyed warm relationships with other aunts; Patty remembers her aunts fondly from childhood experiences and family reunions, especially as sources of family stories and legacies of her heritage.
>
> *(Ellingson & Sotirin, 2010, pp. 9–10)*

This statement reveals key identities and shows Patty and me in relation to each other as co-authors. I acknowledged my positionality differently in my health communication research, describing myself primarily as a female cancer survivor in my study of a geriatric cancer clinic (Ellingson, 1998, 2005). I offered still another explanation of myself in the beginning of my book about crystallization in qualitative research (Ellingson, 2009a), which is quite similar to my embodied description in the introduction to this book, with the exception of a major change in my health status that occurred between publications—from having an impaired leg to being an amputee. As Martinez (2000) noted in a phenomenological study of her own Chicana identity, her experience of race/ethnicity "is not mere chance but is always connected to the general circumstances I share with others who are similarly situated" (p. x), just as Martinez' and my experiences are related to others' experiences of female bodies and mine to those of other people with disabilities. Many other interpretive research exemplars include acknowledgements of researchers' embodied standpoints and the political discourses surrounding identity categories (e.g. Newman, 2011; Scott, 2015).

In the middle ground, qualitative researchers adopt interpretive, social constructionist, or postmodernist-influenced perspectives of meaning as intersubjective and co-created, but also as retaining emphasis on the significance of commonalties and connections as generated in systematic analyses (Ellingson, 2009a). Middle ground work often concerns the construction of patterns – themes, categories, portrayals—and practicalities—applied research and recommendations for action to effect material change (Ellis & Ellingson, 2000). Grounded theory methods are typical of this centrist approach. Grounded theory can be situated within social constructionist theory (Charmaz, 2006), reflecting the relational nature of all knowledge claims; that is, meaning resides not in people or in texts but *among* them. The always changing nature of embodiment allows the exploration of liminal states of being and inconstant bodily states (Gilbert, Ussher & Perz, 2013; Guntram, 2013; Guzzardo et al., 2013; Williams, Weinberg, & Rosenberger, 2013), particularly with health, illness, pain, pregnancy, phantom limbs, and post-surgical bodies (Birk, 2013; Draper, 2003; Kelly, 2009; Montez & Karner, 2005). These middle ground forms of analysis rely almost exclusively on traditional research report format for representation, but some researchers complement reports with more creative ways of sharing findings within and beyond academia by blending art, science, and interpretivism in generative ways (Atkins, 2015; Harter & Hayward, 2010; Kennedy, 2009).

Straddling the Continuum

Traditionally, researchers triangulate by employing multiple quantitative and/or qualitative methods to capture more effectively the truth of a social phenomenon (Lindlof & Taylor, 2010). In positivist and (some) postpositivist research, triangulation or multiple methods design involves an attempt to get closer to the truth by bringing together multiple forms of data and analysis to clarify and enrich a report on a phenomenon (e.g., Creswell & Clark, 2010). Analytical pluralism involves multiple forms of qualitative analyses of the same data set (Clarke, Willis, Barnes, Caddick, Cromby, McDermott, & Wiltshire, 2015). While methods may complement and even contrast in terms of procedures (e.g., interviews and surveys with Likert scales), the epistemological underpinnings generally remain consistent, with both upholding postpositivist goals of generalization and prediction. While such work often includes both qualitative and quantitative data or a range of different qualitative data combined into a single report, manuscripts tend to reflect traditional writing conventions and thus do not address embodiment of researchers or (generally) participants.

Embodiment can be addressed in meaningful ways in paradigm-spanning or multimethod/multigenre research in part by showing how embodiment appears and seemingly disappears between and beyond genres of representation. Innovative ways to blend or transcend art and science in qualitative data collection, analysis, and representation proliferate, with more being developed on the basis of community engagement and research (Harter, Hamel-Lambert, & Millesen, 2011; LeGreco, 2012; LeGreco & Leonard, 2011; LeGreco, Leonard, & Ferrier, 2012), arts-based research (Faulkner, 2016; Leavy, 2015), and crystallization (Ellingson, 2009a). Hybrids may integrate conventional analysis with other methods of representation that more readily highlight issues of embodiment, such as photography in a study of qualitative of life among African American breast cancer survivors (López, Eng, Randall-David, & Robinson, 2005). Another way to overtly blend the voices of art and science is to weave them together. *Layered accounts* move back and forth between academic prose and narrative, poetry, or other art, revealing their constructed nature through the juxtaposition of social science and artistic ways of knowing bodily experiences, such as Rambo Ronai's (1995) classic, deeply embodied, exquisitely painful account of surviving childhood sexual abuse (for other embodied layered accounts, see Ellingson, 1998; Strasser, 2016; Tillman-Healy, 1996; Vannini, Ahluwalia-Lopez, Waskul, & Gottschalk, 2010).

Understanding that embodiment in research practices and products necessarily will vary across the qualitative continuum empowers researchers to make strategic decisions about engaging with their body-selves, participants' body-selves, and relevant actants and discourses. Some practices will remain part of the background of researchers' processes, while others can be made explicit. A range of artistic, interpretive, and postpositivist analyses and representations can come from a single data set, highlighting and more subtly imbuing work with attention to embodiment (Ellingson, 2009a). Choices about embodiment—like any choices about research practices—have ethical implications and opportunities; the next section addresses ethics in all phases of embodied research.

Ethical Bodies in Research

I discuss embodied ethics in terms of three broad phases: research purpose, roles and conduct, and representation, and offer concepts to aid researchers in understanding how to move from a dutiful ethics that researchers fulfill in order to place a check mark in a box on an official form and meet the minimum required standards, to an ethics that researchers enthusiastically *do*, that we practice with overt attention to the material consequences of researchers' ethical choices.

Ethics of Research Purpose

All research is value-laden, and all knowledge produced has the potential to harm as well as benefit (Reinharz, 1992). In research that focuses on embodied issues in topic, practices, or both, the *purpose* of the research is grounded in the material world. That is, the research purpose connects values and goals with explicit and in-depth attention to how participants' material worlds may be constituted through and affected by research processes and outcomes, and how research outcomes and representations may be used in the world. The creation of embodied knowledge is an opportunity to make overt the politics that were implicit or to harness the power of research to move toward praxis (i.e., making a difference in the world) (Preissle, 2007). Knowledge production can foster equality in many ways, including: critiquing dominant, taken-for-granted power relations; focusing on marginalized people's lives and experiences on their own terms (not just in comparison to dominant groups as normative); representing diverse standpoints within scholarly disciplines; embracing research as praxis (fostering social justice), promoting understandings of intersections of gender with other identities, and encouraging participatory research processes where appropriate (Preissle, 2007). Ethical embodied research repositions awareness of bodily needs—e.g., food and transportation, safety and well being, child care and elder care, vulnerability and potential for growth—from an afterthought to a primary concern in all decisions about initiating, conducting, and representing research. In other words, embodied research should be explicitly about understanding, problematizing, critiquing, and addressing material conditions for participants and their communities. Ideally, attention to issues of embodiment should be explicit, rather than implicit or footnoted, in the stated purposes of qualitative research projects; of course, this is not always possible. However, even if research representations cannot—for various strategic, legitimate reasons—overtly refer to embodiment of research and participants, research purposes must include careful consideration of embodiment. Moreover, ethical embodied purpose is only the beginning; the material consequences of research processes must be humanely addressed in order to foster an embodied ethic of social justice.

Ethics of Research Roles and Conduct

Federal and institutional guidelines about researcher conduct toward research participants address bodies in a number of ways. Sometimes this attention is explicit, as in trials of medications that will be administered to participants' bodies, or experiments during which the participants' body-selves may be manipulated and measured. Yet for most qualitative research, attention to bodies in terms of research conduct is at best implicit and at worst ignored. Codes of research conduct require informed consent, risk/benefit analysis, and fairness in selection of participants (Preissle, 2007).

Feminist methodologists favor protection of participants, of course, but posit that such embodied protection is not best ensured by detached, knowing, authoritative researchers who exercise power over subjects, understood as unknowing sources from which data are extracted (Preissle, 2007). Instead, it is through engaged researchers' close encounters with participants that we come to know them and hopefully to act in their best interests. Feminist, poststructuralist, and posthumanist theorists would agree that researchers "don't obtain knowledge by standing outside the world; we know because we are part *of* the world" (Barad, 2007, p. 185, original emphasis). Immersing our body-selves in participants' worlds has serious ethical implications: "knowing is an act of responsibility to the agential cuts I help make as I intra-act with the objects, ancestors, participants, culture, discourses, and so on that constitute my study. I cannot remove myself from that responsibility" (Nordstrom, 2015, p. 396). For example, the ways in which we behave toward interviewees have bodily consequences—do we bring gifts of food or other treats to share, do we accept their offerings of food when we know that they do not have enough for themselves? Do we have sexual relationships with participants while doing ethnography which impacts not only their bodies (e.g., risk of sexually transmitted infections) but also in cross-sex interactions could result in pregnancy (Grauerholz, Barringer, Colyer, Guittar, Hecht, Rayburn, & Swart, 2013; Nelson, 1998)? Do we use our bodies to work alongside participants in their businesses, potentially increasing their access to resources? This is not to dismiss concerns about psychological and emotional aspects of participants' lives, but to shed light on material aspects of participants' well being from an embodied perspective. I share several concepts I have found helpful for thinking through an *embodied ethics* of researcher conduct.

Being-with

Drawing upon the work of Merleau-Ponty (1968) and Ahmed (2000), La Jevic and Springgay (2008) assert that embodied ethics is not merely a set of moral guidelines and rules about behavior toward participants (although those also may be useful). Ethics instead (particularly feminist ethics) must be concerned with "how one can live with what cannot be measured by the regulative force of morality" in our embodied interactions with Others (Ahmed, 2000, p. 138). The authors promote an "ethic of being-with" that embraces the embodied nature of encounters.

> Being-with constitutes the fabric of everyday life and the ethical encounter. Through bodied encounters, body/subjects create lived experiences together and nurture one another's ethical relationality. In other words, all bodies/ subjects involved in the research inquiry are active participants whose meaning making exists in the moment of encounter.
>
> *(La Jevic & Springgay, 2008, p. 70)*

Ethical encounters are not merely abstract but occur on an intersubjective, interpersonal level where body-selves encounter one another and interact in relationally constituted moments, reflecting "an ethics of embodiment [that] is concerned with the processes of encounters, the meaning that is made with, in, and through the body" (La Jevic & Springgay, 2008, p. 70).

To embrace an ethics of being-with, researchers face the necessity of being culturally competent and knowledgeable about vulnerable groups (Smith, 2009). In the bodily encounters, researchers and participants negotiate differences and commonalties in clothing, personal boundaries, touching, modes of greeting, and, at times, languages and issues of translation. Embracing an embodied ethics of being-with participants directs researchers' focus on the embodied intersubjectivity of research processes and of participants' everyday lives.

Compassion

A second concept for embodied ethics is compassion, which reflects a feminist ethic of care and balance of focus on self and other (DeVault, 1990; Preissle, 2007). While researchers typically think of compassion as directed at participants, practicing self compassion can help people to communicate more effectively and more humanely with others (Neff, 2003). Embodiment of compassion involves specific decisions about how to treat others, such that "[e]ach interaction should be fundamentally relational and visibly be an ethical moment of care" (Glass & Ogle, 2012, p. 71). Becoming close to participants is accomplished through bodily performances of caring and compassion, all of which are enacted through the body—e.g., soft tone of voice, touching a hand, hugging, kind facial expressions. An excellent example of embodied compassion in practice described compassionate interviewing with Holocaust survivors, a particularly vulnerable yet resilient group of people (Ellis, 2017; Ellis & Patti, 2014; Patti, 2015). Ellis and Patti present compassion as a holistic mind-body-spirit practice of caring for self and other that involves listening deeply, giving undivided attention, and authentic caring about the other person.

Moreover, caring relationships with participants are not always accomplished through in-person interactions. Online research is common (Fielding, Lee, & Blank, 2008), ethnography with virtual (Internet) communities (Markham, 2005; Parker Webster, & Marques da Silva, 2013), and interviewing through email, instant messaging, and other computer mediated communication (Cook, 2012; James & Busher, 2012). Researchers must remember that embodied vulnerability of participants is just as much a concern with online research as that conducted in-person (Liamputtong, 2007).

Dynamic

Because "an ethics of embodiment is complex and dynamic, open to challenge and revision" (La Jevic & Springgay, 2008, p. 71), no single ethical standard can encompass all situations that may arise during the course of research (Swartz, 2011). This requires researchers to be flexible, attentive, and committed to ongoing consideration of not just embodied well being or material circumstances but even dynamic embodied identities and categories. Bodies may be framed in complex ways in participant selection, for example: race, gender, and age as seemingly straightforward identity categories, or practices such as men who have sex with men, athletes who play particular sports, or people with particular physical diagnoses (e.g., cancer, HIV) or conditions (e.g., pregnancy, muscle development). Recruitment for these categories is generally considered straightforward but may be far from it, as bodies change during the course of research. Embodied ethics includes the responsibility to make sense of participants in ways that honor the complexity and evolving circumstances in which they live and in which they understand their identities. Accounting for flux in participants' and researchers' body-selves will require careful reflexivity to consider the ongoing becoming of participants' body-selves. In addition, attending to the dynamic nature of bodies may involve including participants in research activities (see the Embodied Practices section of this chapter) that help to highlight a plurality of perspectives, embodied voices, and entry points to participants' vibrant worlds, "an ethics of parallax perspectives" (Swartz, 2011, p. 50).

Public and Private Bodies

Embodied ethics responds to the ways in which research inevitably "makes public the private body. We watch and note the body. . . . The boundary between the private body and the public body is . . . an ethical dilemma" (Coffey, 1999, p. 75). There is no formula for calculating how much and which embodied details of participants to reveal, and while some may be quite obviously unethical (e.g., disclosing confidential medical information without explicit permission), ethical descriptions of embodied identities and acts may not be as easily discerned. When researchers conduct ethnography or interviews, the body is implicitly and explicitly included. We detail what people do, and what they do is done as embodied beings. Fieldnotes during ethnography and interviewing, which also may include photos or video recording, focus on bodily details that may not typically be shared with others. When we inscribe embodied details in our data, we negotiate the private/public tension even before we get to the point of shared representations (i.e., publications). Just attending to bodily practice involves rendering bodies vulnerable through our description and inscription (Geertz, 1973). Since no value-free research is possible, inevitably some judgment, even if compassionate, respectful judgment, becomes evident through researchers' language choices and decisions on which details are included and which are excluded. Describing seemingly mundane details of professional practice in healthcare delivery, for example, has implications for both healthcare provider malpractice and patients' nonadherence (i.e., noncompliance) with recommended treatment. Describing teachers' behavior in a classroom has professional and legal implications for the teachers and students. Great care must be taken to respect the dignity and well being of participants, including but not limited to specific details of their bodies and bodily practices (Damianakis & Woodford, 2012). The bottom line is to try to be a good person and to "first, do no harm," which includes negotiating ever-shifting ethical boundaries of privacy (Saldaña, 2014, p. 979).

Reciprocity

For decades, feminist and other researchers focused on social justice have pointed out that researchers should give back to participants, not just use or take from them, and examine carefully how others benefit, if at all, compared to how researchers benefit from the research (Swartz, 2011). Traditionally, reciprocity in qualitative research was framed as a practical necessity; through developing friendly, open relationships with participants, researchers presumably generated better data (Powell & Takayoshi, 2003). Moreover, researchers should give time, resources, or both as a way of thanking participants and acknowledging their time and efforts to collaborate in research processes. In this sense, "[r]eciprocity is a matter of making a fitting and proportional return for the good or ill we receive" (Becker, 2005, p. 18). Advocates of social justice research (Harter, Hamel-Lambert, & Millesen, 2011), feminist methodologists (Hesse-Biber, 2011), and participatory action researchers (Wang, 1999) frame reciprocity as more complex and political, including guidelines such as do no harm to communities, but also to collaborate with participants as equals, speak *with* rather than *for* participants and highlight their voices, acknowledge embodied participants and their material circumstances, critique structural inequities, and develop solutions to problems identified by participants (Preissle, 2007). Moreover, "[r]eciprocity promotes recognition that partners have varying amounts and types of power in different situations and different interests in a specific project—and thus will benefit from different things" (Maiter, Simich, Jacobson, & Wise, 2008, p. 321). Further, reciprocity can include *catalytic validity*, offering tools so the research process "re-orients, focuses, and energizes participants in what Freire (1973) terms 'conscientization,' knowing reality in order to better transform it" (Lather, 1986, p. 67; see also transformational validity, Koelsch, 2013). Examples of reciprocity are thank you gifts for interviewees such as gift cards to local stores (Ellingson, 2011a), financial assistance with school fees for children (Swartz, 2011), paying for meals or drinks while meeting at cafes (Stodulka, 2015), assisting teachers in classrooms (Gallagher, 2011), assisting in the work of the business or organization, much like a intern or apprentice (Edley, 2000), and providing workshops, skills training, or other consulting or supporting services to a group or organization (Diver & Higgins, 2014). The gift of listening to people, especially those who did not have a lot of other people to talk with about issues important to them or who were going through stressful experiences is also a significant form of reciprocity (DeVault, 1990; Ellingson, 2005; Priya, 2010; Swartz, 2011).

Embodied Self-Care

Stress is a deeply embodied phenomenon. Feminist researchers posit that a meaningful, mature ethic of care involves a balance between care for others and for the self, rather than focusing on one or the other (Noddings, 1984; Preissle, 2007). Ethnographers have long written about the dangers of fieldwork and the serious risks of disease, physical and sexual assault, and other vulnerabilities (e.g., Lindlof & Taylor, 2010). Qualitative researchers have an ethical responsibility to attend to their own embodied well being. This is not only a practical or risk management issue but also an ethical one; surely we as researchers owe ourselves a standard of care equal to that which we would extend to our participants.

Researchers may experience significant costs to our health when we engage in intense interactions and discuss sensitive topics, and the resulting psychological strain and distress may manifest in physical symptoms. One research team described reaching "researcher saturation—more commonly referred to as burnout—and gradually emotional overload led to our experience of distress: frequent headaches, anxiety and panic, a 'foggy head,' dizziness, nausea, and even a cyst on the vulva, reinforcing the embodied experience" (Wray, Markovic, & Manderson, 2007, p. 1397). Researchers should endeavor to prepare for the rigors of data gathering by making a plan for self-care throughout the research process, even if it slows down progress. Fieldwork and interviewing can be exhausting, stressful, and even dangerous (Corbin & Morse, 2003; Warr, 2004).

Two quotes have helped me to keep the importance of researcher self-care in mind. One is from creativity guru Julia Cameron (2002) who urged her students to repeat this mantra to themselves: "Treating myself as a precious object will make me strong" (p. 101). The image of a precious object brings to mind gently carrying or safely storing a material item, keeping it clean, and taking steps to ensure it is not harmed—all of which invoke deliberate, physical caretaking. The second quote comes from the Buddha, who is reputed to have said:

> You can search throughout the entire universe for someone who is more deserving of your love and affection than you are yourself, and that person is not to be found anywhere. You yourself, as much as anybody in the entire universe deserve your love and affection.
> (https://www.goodreads.com/author/quotes/2167493.Gautama_Buddha)

Self-care is arguably the most important way of demonstrating self-love and self-respect for researchers. We may compare our own suffering to that of participants and feel unworthy of care given the greater magnitude of participants' suffering. But the Buddha's words remind us that all deserve to receive loving care, and that care needs to be given to our body-selves in material ways, including but not limited to the possibilities listed in the Embodied Research Practices section of this chapter.

Ethics of Representation

The third main area of embodied ethics is the ethics of embodied representation in the whole range of research products. "The ethics of representation is the good or ill that results from how participants are represented in publications, presentations, and other reports of research" (Preissle, 2007, pp. 525–526). Bolen and Adams (2017) refer to the capacity for representing research as "narrative privilege," which they suggest consists of, among other ethical issues, considering who has the (physical and mental) ability and resources (time, money, technology) to write and who is afforded a hearing—and by whom and in what context(s)—and "explicitly acknowledging these concerns—being reflexive about the stories we tell—is an even better, often more credible practice" (p. 623). Narrative privilege exists whether researchers produce traditional academic narratives—i.e., research reports—or (as Bolen and Adams discuss) artistic narratives, e.g., autoethnographic stories, poetry, film, performance, photography, and so on. In general, feminist and other critical theorists encourage researchers to construct representations that support participants' embodied well being, pride (e.g., do not embarrass participants), and intersectionality of embodied identities (e.g., avoid essentializing people through a single identity category) (Preissle, 2007).

Feminist researchers often promote alternative representations—member checking, participatory approaches, multivocal representations—but all options have strengths and limitations, and none is ethically perfect. Creative analytic (Richardson, 2000) or multigenre/multimedia representations of research findings involve just as much sense making as writing conventional research reports and similarly risk othering participants' body-selves (Ellingson, 2009a; Preissle, 2007); that is, experimental texts do not escape their authorship (Haraway, 1988). Stories create and inscribe bodies with meaning through the imposition of order on unordered events (Frank, 1995). While stories invite readers to more directly fashion personalized, multiple meanings, the author's body-self still underlies decisions on what to include and exclude from the narrative. Researchers accomplish representation of others via the author's power, even if that power alters and/or diminishes somewhat through highlighting participants' words, perspectives, and knowledges. Researchers "choose the words, we choose the placement of the words on the page, we choose the moment to capture, we guide the gaze of you, the reader" (Honan, 2014, p. 11). Representation inescapably involves an invocation of authorial power and control as we inevitably speak for others' body-selves, which is a tremendous responsibility.

Practices for Embodying Design

Wondering Bodies

Elsewhere I articulated a process of *wondering* as part of research design and ongoing processes of research, as qualitative research is (in)famously nonlinear with plans, relationships, and outcomes in flux (Ellingson, 2009a, pp. 74–77). I developed a list of questions for initial and ongoing reflection regarding data and forms of analysis, topics with possible (inter)connections, audiences for research findings and implications, researchers' desires regarding themselves and their research, and potential genres for representation. All of those questions remain relevant, and I add a list of questions that focus more explicitly on embodiment issues for ongoing reflections and opening of possibilities.

- To which demographic categories do I generally consider myself to belong? What socially marginalized identities do I claim and how do they manifest in my body? From what forms of cultural privilege do I benefit? In what ways do my privileges and oppressions intersect, and what consequences do these have in my daily life? How are my intersecting identities hidden and revealed in my research processes and products? How do I experience my body on a daily basis? How has my body-self changed throughout my experiences as a researcher, and how has it remained consistent?
- To which demographic categories do my participants belong? What socially marginalized identities do they hold? From what forms of cultural privilege do they benefit? How are participants' bodies pathologized, scrutinized, medicalized, racialized, or otherwise subject to cultural narratives of their group(s) value and status? What important bodily differences manifest within my group of participants make them a heterogeneous group, even with all that they have in common? How do I understand the daily routines in which their bodies engage? What specific bodily practices constitute their work or mission?
- In what ways do my privileges and oppressions intersect with participants' privileges and oppressions, and how do these function in our ongoing relationship? How do our areas of similarity and our areas of difference relate to social, economic, and political power? What unique, embodied experiences of participants do I seek to understand, and how do my own embodied ways of being in the world shape and constrain those understandings?

Free writing, or journaling, on these questions at the early stages of a research project and then periodically thereafter, using writing as a method of inquiry, is a fruitful practice (Richardson, 2000). That is, I suggest that researchers concentrate on answering these questions as a practice, focusing less on finding answers and more on the process of thinking through these important issues. Also, I highly recommend the list of provocative questions for reflection offered by Brady (2011) as suggestions for researchers engaging in what she calls the embodied process of "cooking as inquiry," which absolutely are more broadly applicable to considerations of doing embodiment.

Document Embodied Design/Processes

The processes of research design and planning implicate bodies in complex and dynamic ways, so start a research journal for reflections on how you are defining the bodies of your participants for purposes of recruitment or for negotiating access to a place, how your own body may be implicated with those people, in those spaces, what physically you are proposing to *do* in the field, and your understanding of your own embodiment at the beginning of your project. Ideally, you will be able to capture some of the fleeting moments of embodied knowing and sense making that occur early on in your project, which will be valuable to reflect back on later. Ideally, such reflections inform not only your understanding of the ongoing negotiation of research design but also the focus of your study. One sociologist articulated the link between embodied research design and understanding of his setting (a boxing gym) as an embodied phenomenon, suggesting that:

> The very design of the inquiry forced me to constantly reflect on the suitability of the means of investigation to its ends, on the difference between the practical mastery and the theoretical mastery of a practice, on the gap between sensorial infatuation and analytic comprehension, on the hiatus between the visceral and the mental, the *ethos* and the *logos* of pugilism as well as of sociology.
>
> (Wacquant, 2009, p. 122; original emphasis)

Demystifying qualitative research processes functions as a way of highlighting and further exploring the ambiguities inherent in all qualitative research. Moreover, such accounts constitute a crucial service to new researchers (and to those wondering why their research never seems to go as smoothly as it appears in others' published accounts) who benefit from their predecessors' experiences as maps of the qualitative territory. Eventually such reflections could spark a reflection on your research processes for a methodological journal or the commentary section of a research journal (e.g., Ellingson, 2009a).

Frame Difference as Relational

For those working on research with partners or teams, critical co-constructed autoethnography creates spaces for collaborating researchers to work across differences grounded in bodily categories and identities, such as race and gender. Cann and DeMeulenaere (2012) recommend that researchers write separate narratives, gently question each other about those narratives in a dialogue, compose retrospective fieldnotes on conversations that proved to be insightful, and note breakthroughs or critical moments in sense making. Next, the authors "knit a cohesive piece that intersects with theory and prior literature, makes transparent our theoretical frameworks, makes an argument and uses our narratives (and other research) to buttress that argument" (Cann & DeMeulenaere, 2012, pp. 150–151). Such an approach builds on the strengths of autoethnographic reflection on researchers' participation in their projects to enable comparisons, connections, and questions to be made among or between researchers that interrogate the constitutive role of standpoints in our analysis and writing.

Embrace the Mess

Despite researchers' efforts to contain them, bodies are messy. Methods also are messy in practice, no matter how carefully planned. Rather than trying to avoid and minimize the messiness of bodies intra-acting within a field setting, we can embrace the mess as generative; "the idea of messiness enabled us to imagine together the ethical, the theoretical, the methodological, the experiential, and the emotional. We were, so to speak, *moved to messiness*" (Avner, Bridel, Eales, Glenn, Walker, & Peers, 2014, p. 56; original emphasis). Collaboration between people is neither a linear nor a neat process, especially across disciplines or identities, and feeling disorganized and in a state of continual flux can be discouraging for researchers. Yet messiness also can be generative:

> The variety of disciplinary and theoretical lenses ultimately became an ethics of messiness and multiplicity; the messiness of bodies, the messiness of emotions, and the messiness of human experiences of movement encouraged us to re-think and challenge traditional dichotomies and hierarchical understandings, which potentially over-simplify and close-down our emotional experiences of physical (in)activity. (Avner et al., 2014, p. 61)

The necessity of mess is honored explicitly by indigenous methods which may not assume individualism, linear time, or cause and effect in the same way that mainstream methods do (Madsen, 2015).

Transgress Methodological Boundaries

I encourage researchers to consciously and deliberately "employ a strategy of excess and categorical scandal" (Lather, 1993, p. 677), refuse to play within the lines, make up new rules, transgress categories to offer your own plan and be "promiscuous" (Childers, 2014) in your methodology. By "paying close and careful and critical attention to the assumptions we make" researchers can "remind ourselves of the habits of our practices, to disrupt and interrogate those practices in order to create something different" (Honan, 2014, p. 15). In her study of urban high school students, Childers found her research constituted in and through the material conditions of the school and of the student body, understanding:

> These feminist entanglements as promiscuous, loyally disloyal, and wonderfully infectious. The materiality of fieldwork pushed me to think differently about representational and discursive boundaries circumscribing what counts as feminist research to see the force of a material–discursive feminist inquiry and the agential potential for it beyond gender. . . . . [M]y feminist training has significantly shaped . . . how I engage the world in its complexity, and this way of knowing through being is feminist in its becoming.
> *(Childers, 2013, p. 607)*

Most researchers are trained in a single paradigm, and we may feel we need permission to depart from our mentors' path or to construct multiple parallel or intersecting paths; if you need permission, I hereby grant it to you. Such a becoming approach to methods leaves plenty of room for new ways of bringing theory and method together and of creatively playing with practices in order to be both practical and innovative.

Do Your Embodied Best and Do Not Stop

This caution reflects my most ardent concern: Please do not give up on the project of researching the embodied experiences of others, especially marginalized others, because of fear of exploitation, appropriation, or otherwise unethical representation. I am fully aware of and concur with critiques by feminists and other critical theorists of the dangers of speaking for others, of co-opting their voices, taking their stories out of context, turning stories told into stories analyzed, and exercising our power as researchers in ways that benefit us as academic professionals but benefit our participants little or not at all (e.g., Hesse-Biber & Brooks, 2007; Reinharz, 1992; Roof, 2007). However, we must not let ourselves off the hook of duty and passion by surrendering the practice of researching others. To surrender leaves the study of and with marginalized people to those most vested in the status quo and least interested in transformation. "Your *angst* and *guilt* about your benefits [from representing Others] cannot eclipse or cloud your *responsibility* to do meaningful work" (Madison, 2005, p. 135, original emphasis). Rather, with great care, humility, and good will, we must continue to work with Others in order to build understanding. Doing embodiment in our research foci and processes is not a panacea for the risk of exploitation, but a focus on material conditions and material consequences can keep participants' well being central to our ethical decision making and analyses.

Remember the Care of Body-Self

Researchers need and deserve to care for our body-selves. We need to take proactive steps to project our mental and physical health. Institutions should recognize the need for self-care by researchers (Rager, 2005). It is easy to become accustomed to allowing our bodies' needs to be postponed and to go unmet for long periods of time. While this may work in the short term, over the long haul the cost to researchers' health may be significant. Moreover, such choices may make researchers miserable, and in the end not be nearly as convenient or efficient as planned, as when stress headaches or the flu (caused by chronic stress, fatigue, and lack of self-care) leave a researcher incapacitated for days, making data collection temporarily impossible. Sadly, being over-worked and stressed is considered by many academics to be a badge of honor or even a necessity for a successful career. While I agree that being busy is necessary for success, subjecting our bodies to endless stress and little or no self-care is not only unkind, it is counter productive.

Have a plan for self-care and do not allow time pressures to derail your schedule. Different strategies are helpful for different people, and your own self-care needs and preferences may vary over time and circumstance (Rager, 2005; Warr, 2004; Wray et al., 2007, p. 1399). While some suggestions may seem more body centered and some more psychological, all affect our whole body-selves, just as stress does.

- schedule breaks throughout the day when collecting data
- find peers and possibly a professional to debrief with in order to process thoughts and emotions
- get enough sleep
- eat to meet one's bodily needs
- explore meditation, yoga, tai chi, prayer, or other spiritual practice
- engage in moderate physical exercise
- keep a journal
- spend time with friends, family, pets
- read for pleasure
- spend time outdoors
- listen to some combination of music, audiobooks, podcasts, or meditations

Everyone must find their own way. I have found that my body-self does best with plenty of friends to talk to (some via email), regular riding an exercise bike while reading, my cats, and lots of light reading.

3

INQUIRING BODIES

Embodiment as a Research Focus

Doing Legwork: Languaging Disability in Everyday Life

Glenn, my spouse, gives me a quick kiss as he strides into the kitchen where I am preparing dinner. "Hey sweetie," he says, moving past me to the fridge, where he grabs a Coke Zero and pops it open, taking a long swallow. I heard his car come up the driveway moments ago as he returned from work.

"Hey, love," I reply. "How's it going?" I return to the cheese sauce I am making, stirring more cheese into the thickened milk. I taste it and then resume grating the pungent, extra sharp cheddar that we both prefer, despite its relative lack of convenience compared to the readily available, pre-shredded milder cheeses. As I grate cheese over the pan, I shift uncomfortably, trying to ease the back and hip pain that I typically feel at the end of a day of navigating on my prosthetic leg, trying to more comfortably distribute my weight between my bio-leg and my prosthesis.

"Funniest thing happened at work today." Glenn laughs, his green eyes sparkling. "I was leaving a meeting with Tom, and we were discussing the Sox."

"Mmm hmmm," I respond absently, well aware that Glenn's new boss shares our loyalty to the Boston Red Sox [baseball team in U.S.].

"Well, I was telling him about last year, you know, when we got to sit on top of the Green Monster [left field wall at Fenway Park],' cause he loves Fenway, you know? And I was like, 'yeah, my wife has a disability now, so we get *great* seating at Fenway.' And Tom goes, 'Oh, gosh, I'm so sorry about your wife.' And he looks so serious and uncomfortable." Glenn shakes his head and chuckles. "And I said, 'No, she's fine, don't worry about it. She had her leg amputated last year, but she's fine now.' And I'm trying to tell him about the awesome accessible seats and the view of the outfield, and that homer that Veritek hit right into our section. But Tom's still all flustered and doesn't know what to do since I'm not all worried about you." Glenn shakes his head, grinning.

I laugh. "Yeah, that sounds about right. Most people think it's kinda tragic that you're stuck with a gimpy wife."

"Nah, you're the best," says Glenn, leaning over to kiss the top of my head.

Smiling at him, I nod toward the cabinet to my right. "Grab some plates, would you? Let's eat," I reply.

* * *

"Hey, thanks for meeting me," I say, sliding into a seat opposite where my friend and women's studies colleague Linda is sitting in the front lobby of the student union building.

"Sure. You want some coffee?" she asks, gesturing with her cup.

"Brought my own," I reply, reaching into my bag to pull out another of my ever-present cans of Diet Coke.

We chat for awhile about campus politics and mutual friends and colleagues, and I shake my head, thinking of one kind woman who nonetheless drives me crazy at times. "I saw Ellen again. Every time she sees me, she shakes her head and gets tears in her eyes and says, 'Oh you are just *such* an inspiration.' She's so sweet, and I know she's trying to be supportive, but it's been over a year since I lost my leg. I'm just not feeling very inspiring, you know? Mostly I'm just frustrated. And tired."

Linda laughs. "Yeah. That's you—the wind beneath my wings!" she quips. We laugh, and then she pretends to be worried. "Wait a minute—does that make me the wings above your wind?!"

I laugh, relaxing into her humor, and shake my head.

* * *

Jill moves around her great room, efficiently sorting, cleaning, and organizing everything in her path, as I make "Friday night pizza" for us and her kids, that is, pizza margarita. As I seed a tomato, I can smell the spicy basil, tart tomatoes, and the clean, sweet scent of the fresh mozzarella cheese that I buy each week at a specialty market.

"So I sort of messed up this week," Jill calls out from across the room, jolting me back to the present moment.

"Yeah, how so?" I call back, not lifting my head as I slice tomatoes.

"Well, you know how you call your disabled parking permit your 'crip tag?'"

"Uh huh."

"Well, I was driving around downtown with some ladies," she explains, "and we couldn't find a place to park. We were going to have lunch." Jill pauses to scoop Lego blocks into a bin. "And I said, 'It's too bad we don't have Laura here with her crip tag, we'd be able to find a space much more easily.' And they looked at me, scandalized! I mean, you should have seen their faces."

I snort. "That's my bad—I should have warned you. I like it when you call it a crip tag, because I know that means that you *get* it, that you think my disability is a part of my identity, not something to pity or ignore or be weird about. So you can use the language. But it's kinda like the n-word that only African Americans can say; only people with disabilities are allowed to use 'crip' that way in public."

"But my best friend has a disability, so I thought I could use it." Jill shakes her head. "Clearly not. I meant it in a good way. Obviously."

I chuckle at her dismay. "I know, and you can make crip jokes with me all you want. I hope you will. But yeah, you pretty much have to limit yourself to just teasing *me*, not others."

Jill nods as she continues to cut a wide swath of tidiness across the room.

* * *

Mike, a friend and colleague in my department, lounges in the moss green easy chair in the corner of my office as we discuss campus politics. "What did you say at the open forum?" he asks.

"I explained why it is disrespectful and inappropriate for faculty and staff to have their health insurance coverage be subject to Catholic theology," I say with considerable passion. Lightening my tone a bit, I continue, "I'm putting my *one* remaining foot down on this issue!"

Well used to my gallows humor, Mike laughs. "Yeah, that'll work."

* * *

In this chapter I seek to answer the questions succinctly posed by Perry and Medina (2015, p. 5): "How can we make the body substantive in our research? How do we talk or write about the body?" Theory and research from the rich field of body studies is a great source of suggestions and inspiration for qualitative researchers seeking to do embodiment in a deliberate, reflexive, and creative manner. "[B]ody studies scholars interrogate *embodiment*, or the sociocultural relations that act on individual bodies" (Bobel & Kwan, 2011, p. 2; original emphasis). Examining just a small sample of the research in this interdisciplinary field illuminates many possibilities for troubling the taken-for-grantedness of how we and others *do* our bodies—the many mundane practices that we enact until our bodies are trained to respond automatically (e.g., brushing teeth, Myers, 2008) or conversely, the extensive training dancers undergo to achieve heightened awareness of every minute movement of muscles all over their bodies (Quinlan & Harter, 2010). When researchers attend to our bodies and those of our participants, whether as part of the process of research, the focus of it, or both, our bodies encounter, intra-act, and are always becoming.

I review theory and research on embodied identity and the relationships among identity, demographic categories, and intersectionality. The body as a project, vulnerable bodies, and mundane bodies as modes of doing embodiment are also considered.

Body and Identity

The self is not singular, stable, or unitary (Gergen, 1994). The fixed self is an Enlightenment fiction upheld by patriarchal systems and positivist methodologies, called into question by discursive, historically situated, culturally produced body-selves, and by imagining social and political resistance to oppressive systems and ways of being in a just and humane world (Haraway, 1991, pp. 149–181). Bodies are central to our perceptions of our own identities and cannot be separated from our sense of self (Shilling, 2012; Turner, 2012).

> The term "somatics," comes from soma—the body in its wholeness. From a somatic perspective, we cannot distinguish the self from the body. The characteristics that constitute the self (emotions, actions, beliefs, interactions, perception, ethics, morals, and drive for dignity) all emerge from the physical form . . . Somatics rejects the notion that there is a disembodied, self-contained self that is separate from the life of one's body.
>
> *(Strean, 2011, p. 189)*

Identities are constructed within the sticky web of culture by embodied people and embodied communication among them. Cultural ways of being and knowing shape how body-selves move and are constrained, fed, groomed and clothed, touched by others, and put to any number of other uses. At the same time, our bodily capacities and limitations exert influence on the formation, maintenance, and transformation of culture (Shilling, 2012). Embodied subjectivity is the lived self that qualitative researchers observe, question, interact with, analyze, and reflect on (Sekimoto, 2012).

Identities can be interpreted very differently from one place to another. Take for example, cisgender identities, those people for whom their understanding of their gender identity as a woman or a man aligns with their sex assigned at birth (female or male) and their gender expression (feminine or masculine). Thus women who are labeled by medical establishment as female, and more or less fit within society's (heteronormative) norms for femininity in appearance and manner, likely will feel comfortable with their gender performance within most public places. On the other hand a cisgender performance may not fit well with the expectations of a drag performance club or within a gathering of gender-queer activists. A study of intersex embodiment found that women with "atypical sex development" in adolescence used framing strategies in their daily lives to position their identities as women in relation to normative female embodiment, stressing either their similarity to normative female bodies or emphasizing that everyone differs in some ways and hence it is normal to have some embodied variation in their own manifestations of female embodiment (Guntram, 2013).

Unlike earlier historical periods during which visible signifiers conferred prestige, today the privilege of being unmarked, of having one's positionality obscured as normative, signifies power (Thomson, 1997). Members of marginalized groups (e.g., people of color, LGBTQ community members) recognize that bodies are always political and cannot be separated from the selves that produce knowledge: that is, bodies cannot reflect neutrality but rather are "maps of the relation between power and identity" (Rose, 1999, p. 361). Identity categories for researchers and participants are key—but by no means the only—starting points for considering embodiment. I take a closer look at some of the many embodied "differences that make a difference" in the following sections.

Categorical Bodies

Systems of identity categories are developed, imposed, adapted, and resisted as they come to matter in the world. Researchers will encounter

> the pragmatics of the invisible forces of categories and standards in the modern built world, especially the modern information technology world. . . . Each standard and each category valorizes some point of view and silences another. This is not inherently a bad thing—indeed it is inescapable. But it *is* an ethical choice, and as such it is dangerous—not bad, but dangerous.
>
> (Bowker & Star, 2000, pp. 5–6; *original emphasis*)

Too often researchers naturalize categories instead of pausing to consider what categories mean (and to whom) or what they *do*. Researchers taking embodiment seriously need to question the connections between socially constructed identity categories and our body-selves (Sekimoto, 2012). Researchers should engage reflexively with embodied identities "through the constant questioning of the categories and techniques of sociological analysis and of the relationship to the world they presuppose . . . at every stage in the investigation" (Wacquant, 2009, pp. 121). In particular, we should attend to the ways in which embodied being within a cultural system of categories involves making sense of the world from specific standpoints that are always already implicated in complex power relations. Articulating one's standpoint and being conscious of how it shapes one's understanding of a social issue takes significant effort. Power makes some identities devalued in favor of others: "[g]ender, race, or class consciousness is an achievement forced on us by the terrible historical experience of the contradictory social realities of patriarchy, colonialism, and capitalism" (Haraway, 1991, p. 152). Rather than a one-time achievement, these and other embodied identities are lived (or achieved) in an ongoing state of becoming within specific contexts and places, at particular moments of time. Identity categories operate on cultural, institutional, community, and individual levels; race relations for example, are grounded in histories of privilege and oppression, discourses of discrimination, and racial violence through colonialism, slavery, and racist policies and sociopolitical practices.

Embodied categories persist in virtual spaces of the Internet and cellular devices, as well, where they are both obscured and still in play in the processes of research.

> The absence of visual information about the [virtual or online] partici-
> pant functions more paradoxically than one might realize. Socioeconomic
> markers such as body type, gender, race, and class are used consciously
> or unconsciously by researchers to make sense of participants in physical
> settings. Online, these frames are still used but without visual information,
> they function invisibly.
>
> *(Markham, 2005, p. 799)*

Researchers' body-selves in online environments retain the privilege of being understood as capable of knowledge production, while equally (dis)embodied participants are routinely reduced to data sources (Markham, 2005). Identity categorization is not merely social construction but has material effects on bodies, whether they conform to or defy the parameters of their categories. Complicating—and energizing—the matter of embodied identity further is the concept of intersectionality.

Intersectional Bodies

Identity categories do not exist in isolation but within intersecting oppressions and privileges. Drawing on Crenshaw (1991), Collins (2002) noted that the material circumstances of black women's lives differ from those of black men and white women, and suggested that it is at the *intersection* of categories that black women live their lives. "'Intersectionality' refers to the interaction between gender, race, and other categories of difference in individual lives, social practices, institutional arrangements, and cultural ideologies and the outcomes of these interactions in terms of power" (Davis, 2008, p. 68). Intersections are places of activity and interface where structures, discourses, bodies, and actants come together, and often form a site of struggle. Thus women of color do not simply embody race or gender hierarchies; instead, "the raced and gendered body is the intersection of multiple discourses and structures of oppression: She is the point at which racism and sexism collide" (Sekimoto, 2012, p. 234; see also Allen, 2010).

The burgeoning field of feminist neuroscience examines how the brain interacts with and is impacted by social structures, especially in regard to race, class,

gender, sexuality, and disability, as well as how those social structures shape neuroscientific knowledge. Feminist neuroscientist Pitts-Taylor (2016) explores the embodied mind and the "embrained" body, challenging the nature/culture dichotomy and exploring the biosociality of the brain. She argues that identity categories typically are used in neuroscience research as though they were unproblematic groupings rather than social constructions that intersect in complex ways. "Intersectional perspectives help to outline the entanglements of categories in neuroscientific research, for example, when racism and ageism are connected with gender" (Schmitz & Höppner, 2014, p. 3). Heteronormativity, dualistic views of sex/gender, racism, and ageism produce flawed neuroscience research based on assumptions about bodily identity categories that are unsupported by research, overemphasize differences and ignore similarities, and are interpreted through a cultural lens unreflectively utilized by neuroscientists (Schmitz & Höppner, 2014. Such approaches are widely critiqued by feminist and queer scholars with attention to both culture and to the materiality of brains and the nervous system. Neurofeminist "analyses also show how the newest scientific findings are referenced in order to serve the legitimization of social hierarchies, inclusions, and exclusions" (Schmitz & Höppner, 2014, p. 4). That is, how mainstream neuroscience research serves to maintain and naturalize status quo power relations as inevitable and normative under the guise of unquestionable scientific findings.

Ideally, attention to intersectionality resists essentialism in research, or the reductionism of complex, embodied identities to a singular, unified category (Pitts-Taylor, 2015). Intersectionality "encourages complexity, stimulates creativity, and avoids premature closure, tantalizing feminist scholars to raise new questions and explore uncharted territory" (Davis, 2008, p. 79). Focusing on the ways in which bodies are always embodying multiple categories and identities can open up new questions, as with the conceptualization of queerness of identity as "the open mesh of possibilities, gaps, overlaps, dissonances and resonances, lapses and excesses of meaning when the constituent elements of anyone's gender, of anyone's sexuality [or other identities] aren't made (or can't be made) to signify monolithically" (Sedgwick, 1993, p. 8).

I now discuss examples of a variety of intersectional identities, focused on how researchers have explored the intersections of their own embodiment with those of participants in order to shed light on embodiment as it relates to the performance of identities, work practices, experiences of bodily capacities such as pregnancy and athletic ability, bodies with disabilities, bodies rendered vulnerable through illness, and mundane bodies in everyday life. In each, I explore embodied being in the world in order to demonstrate some of the endless potential for attending to bodies as integral to the processes and products of qualitative research. Further examples of demographic/intersectional identity categories are provided in Table 3.1.

TABLE 3.1 Intersectionality and Embodied Identity Exemplars

Identities	Exemplars
Gender/ Women	• Resistance to motherhood myths about infancy and breastfeeding (Faulkner, 2014) • "Abstinence only" sex education disproportionately burdens young women, denies female desire (Fine & McClelland, 2006) • Wearing a "skort" (or running skirt) while exercising or sports (Flanagan, 2014) • Women's embodied experience of their scars following breast cancer surgery (Slatman, Halsema, & Meershoek, 2015) • Women's roller derby, highly aggressive, even violent competitors, with hyper feminine attire (Finley, 2010; Peluso, 2011)
Gender/Men	• Bio-politics of white male privilege (Bunds, 2014) • Father's embodied caregiving and gendered division of domestic tasks during infant's first year (Doucet, 2009) • Digital "selfies" pictures used by Spanish men in online dating sites and other social media (Lasén & García, 2015) • Midlife alcohol consumption as gendered self-regulation based upon years of drinking (Lyons, Emslie, & Hunt, 2014) • White masculinity performance through heavy alcohol consumption among college age males (Peralta, 2007)
LGBTQIA	• Navigating public restrooms by genderqueer and transgender people (Connell, 2011) • Pathologization of intersex people within Christian theology (Hiebert & Hiebert, 2015) • Embodied boundary work in a lesbian niche dating site (Hightower, 2015) • LGBT musicians making music together in everyday life (Miyake, 2013) • Trans people in U.S. sex-segregated prison system (Rosenberg & Oswin, 2015)
Race/Ethnicity	• Race, gender, and sexuality in prom night expectations, preparations, and experiences (Best, 2000) • Everyday experiences of racism (Boylorn, 2011) • Micro-level performance of black and brown bodies in hip hop dance as embodiments of macro-level social inequalities (Roberts, 2013) • Black women's embodied communicative resistance to stereotypes in predominantly white institutions (Scott, 2013) • Korean adoptees' embodied racial and ethnic identities (Walton, 2015)
Socioeconomic Class	• Economic inequality, racism, and the myth of the strong black woman (Beauboeuf-Lafontant, 2009) • Young girls' embodied experiences of femininity and social class (Francombe-Webb & Silk, 2016)

	• Shame, relationality, and the lived experience of social class and gender in higher education (Loveday, 2015)
	• Teen girls, working-class femininity, and resistance in educational contexts (Ringrose & Renold, 2014)
	• Abusive working conditions and deaths of Mexican migrant farm workers (Smith-Nonini, 2011)
Disability	• Body image of amputees (Crawford, 2014)
	• Embodied trauma and disability in father/son relationship (Lindeman, 2012)
	• The embodiment of assistive devices, from wheelchair to exoskeleton (Pazzaglia & Molinari, 2016)
	• Embodied experience of habilitation staff in encounters with children with disabilities (Råsmark, Richt, & Rudebeck, 2014)
	• Men, spinal cord injury, memories, and the narrative performance of pain (Sparkes & Smith, 2008)

Critical race theory explores the ongoing social construction of racial identities and cultures (e.g., West, 2001). However, the analysis of discourse through which those meanings are negotiated does not stop at the level of cultural discourse.

> [R]ace does not simply categorize and label different bodies. Rather, it constitutes the embodied perception through which one sees, moves, and interacts with others. Thus, what slips through the discussion of identity as socially constructed is the analysis of reality as an embodied experience and identity as a situated embodiment of historically ingrained ideologies.
>
> *(Sekimoto, 2012, p. 231)*

Attending to embodied experiences of race is vital for qualitative researchers seeking to understand how body-selves intra-act in the world. For example ethnic and racial identities intersect with religion in an ethnographic study of Mexican American Catholics in South Phoenix, for whom "religion was deeply embodied. Religion was lived and experienced in, by, and through bodies in the streets, in church, and at the [Virgin Mary] shrine" in a local woman's yard (Nabhan-Warren, 2011, p. 379). The Catholic ritual of saying the rosary, for example, involves kneeling, holding a beaded chain with a cross on it, a position of supplication and devotion that the researcher would not have understood if she had not knelt and felt the beads in her hands. Research on responses to spiritual and religious plurality in Canadian health care settings that provided care for ethnic minority immigrants differed in focus, yet reinforced a similar message about religion as embodied practices, not merely a set of beliefs, through which religion had meaning and was experienced by its followers (Sharma, Reimer-Kirkham, & Cochrane, 2009).

Gender is another set of discursive and material performances. Consider two examples of female identity as related to sex and reproductive choices. One study explored the embodiment of intersecting social inequalities – sexism, racism, and poverty—as played out through use of medical and nonmedical contraceptives by poor women in Brazil. These women resisted biomedical discourses that framed them as irrational and irresponsible and persisted in making choices about their bodies and their own goals (De Zordo, 2012). The cultural construction of norms and expectations for female sexuality differs significantly in a study of (in)visibility in the performance of women's orgasms. With female sexuality viewed so ambivalently in Western cultures, a great deal of "cultural meaning loaded onto the presence of orgasm, the necessity of producing an orgasm which is see-able, and documenting the bodily signs by which orgasm can be definitively 'read off' from women's bodies, becomes a cultural preoccupation" (Frith, 2015, p. 387). Female orgasm thus becomes an interior process made exterior for the purpose of cultural validation in ways more likely to serve men's interests (and the porn industry and capitalism more broadly) than the interests of the women whose orgasms are at stake (Frith, 2015).

Masculinity is likewise performed through embodied practices that intersect with age and class (among other embodied identities). For example, deckhands ("deckies") on Australian shark fishing boats

> embody the practical, productive tasks that denote their role. . . . [W]hile many of the tasks for which deckies are responsible are articulable, the ability to perform them skillfully, properly, lies beneath the level of discourse . . . the relationship between young deckies and their skippers . . . is embodied in the flow of learned, physical movement between the two.
>
> *(King, 2007, p. 539)*

Embodiment of masculinity and gendered subjectivities within gendered occupations of firefighting, hairdressing, and estate agency "are generated through men's embodied engagement with prevailing body-based masculine stereotypes and sometimes highly local processes of inter-subjective negotiation and resistance," particularly as masculinity intersects with aging bodies and class status (Hall, Hockey, & Robinson, 2007, p. 550). In a number of different occupations, men both resist and conform to masculine norms, continuing to strive to be understood as masculine, even though they are aware that they cannot fully achieve embodied masculine ideals.

Class identity intersects with race and gender in complex ways (e.g., hooks, 2000; Reay, 1997). Class is illuminated in a study of whiteness and masculinity in the white sport cultures (college football and stock car racing) of the American South (Newman, 2011). Newman uses his own roots in white working-class and poor communities to participate fluently in sexist and racist discourses of these sporting events, despite his current discomfort with such cultures and discourses, demonstrating the possibilities of fluid insider/outsider identities for qualitative researchers. Saldaña's (2014) witty and insightful "blue-collar rant" turns a class lens on the field of qualitative methodology and suggests that researchers might "need to take it down a notch" so that we engage with people and ideas in a pragmatic and more inclusive manner. Drawing on hands-on wisdom of his working-class roots and the orality of embodied speech and nonstandard English, he likens overemphasis on theorizing, jargon, and excessive appeals to exalted experts (e.g., Foucault) to pretention based on upper class social norms. Saldaña critiques those who consider their way to be the only way and who disparage the more "down and dirty" work involved with qualitative research that wrestles with material problems and works directly with people facing material challenges (e.g., homelessness, racism, poverty, lack of health care; see Drew, Mills, & Gassaway, 2007; Finley, 2015). Saldaña urges researchers to resist unhelpful labels, needlessly jargon-laden prose, and theoretical arguments that are so far abstracted from everyday embodied experience that they seem to exist primarily as mind games for the scholarly elite, like "some dumb ol' hound dog chasin' its own tail 'round and 'round and 'round. It *looks* like fun, but he ain't really gittin' anywhere, is he?" (p. 977).[1]

Disability as a category is not based on a set of "predictable and observable traits" but on any deviation from what is culturally determined to be normal bodily appearance, comportment, and movement (Thomson, 1997, p. 24). Disability is yet another identity that we *do* as we do our bodies. To attempt to hide one's disability is also a way of doing bodies: "Passing is an embodied communicative power struggle" (Scott, 2015, p. 241).

> Able-bodiedness is accordingly not an immanent feature of "the body" (as if it could be decoupled from its environment) but is a dynamic index of architectural, economic, industrial, biomedical, discursive, material, informational, affective, political, and sociocultural assemblages. More specifically, able-bodiedness identifies the congruence of these networks with putatively "normal" bodies.
>
> *(St. Pierre, 2015, p. 340)*

Scott (2015) examined stories of women who described the negotiation of femininity and disability in everyday life. Their performances of self illuminate the complexities of physically disabled bodies that can almost pass as "normal" female and the risks of attempting to pass as nondisabled and female. An imaginative study of embodiment of ability and disability in an integrated dance group for people with and without disabilities illuminated the complexity of movement, disability, and aesthetics in an arts-based, nonprofit organization (Quinlan, 2010; Quinlan & Harter, 2010; Quinlan & Bates, 2014). Cognitive and socioemotional disabilities also question what is considered normal and who benefits from such determinations. The current field of mad studies reclaims madness from biomedical definitions of mental illness, offers critical reflections on neurodiversity that ask what purposes are served by the labeling of people with Autism Spectrum Disorder, and explores psychiatric oppression (i.e., diagnoses and medication) as material and embodied aspects of disability (e.g., LeFrançois, Menzies, & Reaume, 2013; Nabbali, 2009).

Embodiment of nonheteronormative identities (lesbian, gay, bisexual, transgender, queer, intersex, asexual, or LGBTQIA) is another key aspect of identity explored in one close examination of a personal journey of transition for a FTM (female to male) transgender person.

> Claiming humanity in my monstrosity as a transsexual, I make my monstrosity human. Charting my socioemotional gendered transitions on my gender journey, I expose others' marginalizing actions, I question the locatedness of positionalities, and I make my many selves legible. I aim to build connections across difference . . . for cisgender people and for shape-shifters alike, so that we might see ourselves in each other.
>
> (Nordmarken, 2014, p. 38)

Fluidity of sex and gender identities raises questions about taken-for-grantedness of intersections of gender, sexuality, and identity. Material bodies, social discourse, and individual performances of identity categories also intersect with other types of embodied experiences, such as efforts to adorn or shape the body, pregnancy, illness, and trauma. In the next section, I discuss a sample of embodied experiences, with further examples available in Table 3.2, including body projects, vulnerable bodies, mundane bodies, and traumatized bodies.

Body Projects

Contemporary bodies and identities (in particular cultures and places) may become body projects. Many people now greatly concern themselves with determining the health, shape, and appearance of their bodies as a central way of expressing individual identities as well as their identity as belonging to a particular religion, culture, or group (Jackson & Scott, 2010). Striving toward bodily perfection is a high achievement of global capitalism and consumer culture. A high degree of awareness that alternative choices for bodily performance could be made, at least by those with sufficient resources, is widespread. At the same time, globalization, economic inequalities, and cultural change have "the potential to throw into radical doubt our knowledge of what bodies are and how we should control them" (Shilling, 2012, p. 5).

The body is a project that can be worked on through choice and effort to shape the self, rendering our bodies "malleable entities to be shaped and honed by the vigilance and hard work of their owners" (Shilling, 2012, p. 7). Ironically, the more complex and anxiety-provoking our fast-paced world becomes, the

TABLE 3.2 Selected Exemplars: Body Projects, Vulnerable Bodies, Mundane Bodies

Topic	Exemplars
Health and Illness	• Metaphors of cancer in the body as re-imagined taking a dynamic view of embodiment rather than a more static one (El Refaie, 2014) • Living with hepatitis C and recovering from heroin use, studying others in these situations (Harris, 2009, 2012, 2015) • Sociability as an embodied dimension of selfhood in people with dementia in long-term care facilities (Kontos, 2011, p. 330) • Traces both the science behind the need for hydration and the marketing of bottled water to consumers that builds on the current trend for consumers to fashion their bodies through biomedical techniques of the self (Race, 2012) • Childbirth experiences as they reflect, resist, and negotiate between biomedical and natural birth discourses (Walsh, 2010)
Everyday Life	• Laundry practices as sensory experiences and the production of gender in the home (Pink, 2007) • Everyday practices related to water and energy consumption (Browne, 2016) • Embodied self is performed relationally through foodmaking (Brady, 2011) • Walking the streets in urban spaces, engaging with the social world (Bairner, 2011) • Embodiment of dance and disability on reality TV (Quinlan & Bates, 2008)
Workplaces	• Menopausal women and hot flashes at work (Dillaway, 2011) • Post-accident identity; abrupt altering of bodies changed daily performances of self as professional in workplace (Scott, 2012) • Breastfeeding mothers in workplaces (Turner & Norwood, 2013) • Embodied shame by staff in higher education, based on experiences of classism and exclusion (Loveday, 2015) • Experiences of everyday smells in offices (Riach & Warren, 2015)
Body Projects	• Female "natural" bodybuilders (Garratt, 2015) • Marking on bodies through tattoos, piercings, and "self-cutting" (Inckle, 2007) • Male athletes' identity, bodies, ageing, and masculinity (Phoenix & Sparkes, 2007; Sparkes, 2012, 2015; Sparkes & Smith, 2002) • Repeated plastic surgeries (Pitts 2003, 2007) • Anorexia as embodied lifestyle choice for beauty and health (Richardson & Cherry, 2011)
Bodies in Crisis	• Rape survivors (Brison, 1997; Minge, 2007) • Survivors of a catastrophic earthquake in India (Priya, 2010) • Abject embodiment of terminal cancer patients in a palliative care unit (Waskul & van der Riet, 2002) • Social rejection of a gender-ambiguous body (Nordmarken, 2014) • Gendered perceptions of violence and trauma in conflict zones (Keeler, 2012)

more it makes sense to people to assert control by closely regulating their bodies (Bordo, 2004). Indeed, "there is tremendous pressure to play by the rules . . . There is no shortage of rules dictating what we should or should not wear, inhale, and ingest; the size, shape, and overall appearance of our bodies; and even our gestures, gait, and posture" (Bobel & Kwan, 2011, p. 1). Body projects cause bodily anxiety and disorders, and incure risks inherent in seeking to control and change, such as the risks of surgical procedures or medications. Moreover, the risk is high for "images of the desirable body to get harnessed to pre-existing social inequities" that favor already privileged groups particularly in terms of gender, class, and white privilege (Shilling, 2012, p. 10).

Despite one's efforts to manage her or his own body project toward specific goals, however, an individual

> simply cannot intend or dictate the discourses or cultural meanings with which her particular speech and acts will collide and collude. The subject may cite and be cited with discourses that she does not desire, intend, realize, or even know of.
>
> *(Dosekun, 2015, p. 438)*

In her analysis of urban African women who affect a hyper feminine style of dress, grooming, and adornment, Dosekun (2015) pointed out that their appearances can be read in a variety of ways by different groups, regardless of the women's intentions. Cultural discourse surrounding femininity is "excitable," which means that its effects cannot be pre-determined or controlled, and its meanings may be co-opted by any number of competing and intersecting discourses, such as class, race, urbanization, and (post)colonialism (Butler, 1997).

Mundane Bodies

Bodies perform continuously in the most mundane of daily activities. Daily tasks require "embodied sensory knowing," which in turn supports the various identity practices by which a particular self is contrived, and including its particular sense of home, among other places (Pink, 2009, pp. 52–53). Emplacement refers to the ways in which embodiment is experienced not only in specific bodies but in bodies as they are present within specific places, such as homes, workplaces, stores, and other public places (Pink, 2009).

> Both home and identity are produced in a set of sensory practices that enmesh or embed the body in place, establishing relations between bodies and sites, even though these practices and relations are not always the subject of conscious reflection. The body acts in its senses, generating a kind of corporeal knowledge that differs from cognition or intentionality.
>
> *(Duff, 2012, p. 273)*

Pink's (2005) video ethnography of how laundry is accomplished examined varying approaches to the sensory experiences of clean and dirty laundry as part of one's experience of home. What constitutes home or meanings of home is continually in flux. For example, home was constructed by LGBT youth after they "come out" to parents and established outside homes for themselves (Pilkey, 2013). The youth often engaged in "productive nostalgia" and sought to recreate elements of their parents' homes for themselves in their lives as adults who no longer hid their sexual identities. Other everyday practices in the home relate to issues such as water and energy resource consumption (Browne, 2016) or food preparation, including one study of how "the embodied self is performed relationally through foodmaking" (Brady, 2011, p. 322).

Everyday life occurs outside homes as well, as body-selves walk the streets in urban spaces while engaging with the social world (Bairner, 2011) or engage in physical activity (e.g., sports and dance) and that includes experiencing embodied emotions (Avner, Bridel, Eales, Glenn, Walker, & Peers, 2014). Likewise, teachers, children, and administrative/support staff in educational settings engage in embodied routines (e.g., Gallagher, 2011). Adults (and some children) spend a lot of their time at work, where mind–body dualism is a "learnt state of being" (Baxter & Hughes, 2004, p. 364). One study challenges this dichotomy by "highlight[ing] how exploring smell as an embodied experience can inform us of how micro-social processes of being together at work are enacted between bodies-who-organize and constitute the conditions of bodily integrity" in organizations (Riach & Warren, 2015, p. 796).

Body-selves also interact in everyday life with vast amounts of technology. "We are our bodies—but in that very basic notion one also discovers that our bodies have an amazing plasticity and polymorphism that is often brought out precisely in our relations with technologies. We are bodies in technologies" (Ihde, 2002, p. 137). An obvious example of this blurring of the lines between bodies and technologies is the use of limb prostheses by amputees.

> [S]uch prostheses contest our faith in corporeal integrity even as they are intended to restore the clean and proper body. They not only demonstrate the inherent plasticity of the body, but, in the very process of incorporating non-self matter, point to the multiple possibilities of co-corporeality, where bodies are not just contiguous and mutually reliant but entwined with one another.
>
> *(Shildrick, 2015, p. 16)*

Beyond this illustrative use of disability, less obvious examples of how technology has become part of our bodily being in the world proliferate. One provocative example of how our bodies work with computers is bodily engagement with Internet pornography.

The computer arranges the viewer's body when browsing pornography online in several ways—for example, the location of feet placed beneath the desk or above the desk, crossed legs, straight legs, curling toes, relaxing toes, sitting forward, sitting back, sitting up, slouching, contracting muscles, relaxing muscles—each accompanied by a whole range of sensations—frustration, pleasure, anxiety, disgust, exhaustion, shame, excitement, panic, and so on. One's body is limited to a number of physical arrangements, confined by the physical presence and specificity of the computer to a particular place.

(Keilty, 2016, p. 68)

Pornography is typically part of the private sphere, but computer and Internet technology also influences our experience (and use) of public spaces (e.g., Farman, 2015).

Vulnerable Bodies

Embodiment of illness and social inequalities (White, 2012) and that of health over the life span (Hockey & James, 2012) are areas of central concern in body studies. Leder (1990) argues that our bodies are normally an absent presence; when they are functioning well, they recede from consciousness, and we use them automatically, without reflecting on how to walk, for example, or noticing our breathing. When something goes wrong, our taken-for-granted bodies make a "dys-appearance" or reappearance, meaning that we become very much aware of them, or at least parts of them (Leder, 1990, p. 84). Pain is a key example of this phenomenon; we find it difficult to ignore areas of our body that are painful (Scarry, 1985). Studies examined the lived experience of chronic pain and embodied identity (Honkasalo, 2001) and the intersubjective nature of pain and the role of language in conceptualization of phantom pain (Crawford, 2009). An autoethnography detailed the challenges of living with invisible chronic pain and trying to make others who downplay or deny the pain associated with invisible disability understand that the pain exists (Birk, 2013). Birk lives with acute exertional compartment syndrome and has survived more than a dozen surgeries, countless hospitalizations and procedures, and years on crutches. Health care providers, colleagues, community members, and strangers have refused to acknowledge her suffering. Birk is not alone; "chronic physical pain—despite its traditionally being seen as the most private and personal of experiences—is also a public, even political issue . . . . in the sense that bodies in pain represent a potential site of social critique and resistance" as sufferers fight for legitimization of their suffering (Birk, 2013, p. 391). Performance of credibility seeks to manage stigma and identity, yet the body-selves of chronic pain sufferers are diminished by the struggle of the pain and the struggle of the alienation/delegitimization of this embodied experience.

Leder's (1990) perspective is useful, but the suggestion that the body can typically be ignored in everyday life (except in case of dys-function) is limited by a privileged standpoint from a white, male, affluent, and presumably heteronormative, nondisabled body-self for whom typical embodiment is considered nonproblematic by social norms. Some bodies enjoy far less capacity for normative receding from consciousness because they are subjected to embodied intersections of racism, sexism, classism, homophobia, Islamophobia, or any number of other forms of discrimination based upon perception of ongoing bodily characteristics (e.g., skin color, breasts), processes (e.g., menstruation), or practices (e.g., wearing a hijab as a Muslim, using crutches or a wheelchair for mobility) (e.g., Crenshaw, 1991; Inckle, 2014).

Messy ill bodies also are stigmatized; a key part of being considered a functioning adult is keeping one's bodily fluids and odors under control (Frank, 1995). Research on those living with IBD (inflammatory bowl disease) reveals a sense of self and relationships with an embodied stigmatized condition that centers on problematic defecation (i.e., constipation, diarrhea, lack of control of elimination, flatulence) in daily life and the embodied communication strategies used to disclose or hide their illness (Defenbaugh, 2011, 2013). Another problematic messiness is the dying cancer patients' abject body, which exists in

> a state in which coherent bodily boundaries erode and the self has little control over the leaking of blood, urine, feces, vomit, bile, pus, and various other hideous body fluids. Its untidiness violates not only biological but also normative boundaries. What is ordinarily inside now comes out, not only threatening the concretion of the body but also resulting in an ominous seepage of matter of physical, personal, moral, and social significance.
>
> *(Waskul & van der Riet, 2002, p. 487)*

Messy bodies and bodies out of control are stigmatized and avoided, and yet those bodies and their boundaries are actively managed by those who are stigmatized.

Likewise, fat bodies are inherently suspect in contemporary societies where they are stigmatized and pathologized in a culture of thinness as both a beauty and a health ideal (Satinsky & Ingraham, 2014; Warin & Gunson, 2013). The "body becoming" theory of fat imagines a variety of responses to fat bodies that are not pathologizing (Rice, 2015). Within fat studies, most intersectional research explores issues of body size and gender and/or sexual orientation, with insufficient attention to other identifies such as race and age (Pausé, 2014). Another type of vulnerable bodies includes the poor women who engage in embodied labor as commercial pregnancy surrogates in India, carrying and birthing fetuses primarily for affluent white women in developed nations (Pande, 2010). Commercial surrogacy in the Third World is just one example of "a continuum of body exploitation at the core of globalizing processes" from pilfering the best athletes to play in affluent countries to "mail-order brides" and human trafficking (Shilling, 2012, p. 22).

People who have been traumatized are yet another form of vulnerable bodies. Trauma is felt throughout the body, both directly through sustained injuries as the result of natural or human-made disasters or individual incidents of violence or tragedy, or through witnessing the same. One researcher studied survivors of a deadly earthquake in India and their embodied trauma and healing, including the researcher's role as empathetic witness (Priya, 2010), while another study highlighted disaster relief workers' resiliency and experiences of providing assistance (Agarwal & Buzzanell, 2015). Another compelling text explores body and memory of rape through "embodied art," which Minge (2007) defines as "the process of inquiry that adds depth of emotion, perception, sensory detail, meaning, and creation of meaning to both the lived and the recreated experience through the blending of various art forms" (p. 252). The autoethnographic text weaves poetry, narrative, and painting to illuminate the embodied nature of traumatic memory. War also induces embodied trauma; one manifestation is the (mis)treatment of women in Iraq diagnosed with "hysteria" which

> often involv[es] brutal acts of violence on the bodies of female sufferers, [and] signifies a denial of female agency, the normalization of particular forms of violence, and the privileging and even legitimization of those forms in the realm of medical knowledge and biotechnology. . . . I seek to show how nationalism, bioscience, and neoliberal narratives of the post-conflict zone coalesce, circumscribing the available means of agency and social experience for women in Iraq.
>
> *(Keeler, 2012, p. 133)*

Traumatized body-selves resist the body–mind dichotomy with embodied memories, visceral emotions, and fleshy responses.

Vulnerable bodies of all types speak not only in words but through the senses. Survivors suffer not only from physical violence and harm, but loss, psychological abuse, and social exclusion. Yet vulnerable body-selves are rarely passive and still resist; "humans can be at once rule-bound and wonderfully inventive agents of social change . . . Every action thus potentially contains elements of both resistance and accommodation" (Bobel & Kwan, 2011, p. 2). By embracing and reflecting on embodied identities and experiences, people with vulnerable or traumatized bodies, and potentially even those engaging in everyday bodily routines or (problematic) body projects, "are better able to imagine alternative possibilities to story ourselves in other, perhaps more fulfilling, relational, socially just, holistic ways" (Brady, 2011, p. 324).

Practices for Inquiring into Embodiment

Making Sense of Stylized Acts

Butler (1990) refers to gender and other identities as performed and constituted through "a stylized repetition of acts" whose meanings are specific to particular sociohistorical contexts. Select an identity being performed by one or more of your participants and list as many stylized acts as you can that contribute to your perception of that identity. The performance could be related to gender, race, age, or other demographic identity category, or it could be related instead to embodied experiences such as training for an athletic event, being a member of a team, performing a professional role or that of a low-status employee in an organization, a fan of a particular music group or a sports team, or other way of being. List clothing and accessories, grooming (e.g., hair), gestures, posture, physical orientation toward others (side-by-side, around a table, at customer service stations, etc.), use of mediating communication technologies (e.g., email, phone and cell phones, texting, instant messaging, mobile apps), and other practices that make up the repeated "stylized act."

Consuming Bodies

Consider how many products and services were utilized by participants in order to achieve their surfaces (Dosekun, 2015; Shilling, 2012). Observe and ask about how their appearances are achieved, the types of efforts that go into hair styles, make-up, nail grooming, clothing acquisition, outfit assembly, selection of accessories, and how they relate to economic resources and gender. Consider the role of actants that enable the achievement of bodies and certain performances of them—hair dryers, hair styling aids, lipstick and other forms of make-up, skin care products, cosmetic surgery or treatments (e.g., Botox injections), hair removal (eye brows, face, limbs, genitalia). What do participants carry their possessions in when they are outside their homes, if anything—pockets in clothing, purse, briefcase, knapsack/backpack, lunch bag, tote, computer bag, satchel, rolling case, disposable bag? How do they interact with their belongings in public, and how does this reflect or contradict their identities? Reflect on your participants as consumers as that role intersects with their embodiment of your topic focus.

Resist "The Body," Embrace Bodies

I have endeavored to discuss "the body" only when I am truly talking in the abstract about the experience of being a body, and even then, I have tried to eliminate most of these references because I concur with feminist theorists who maintain that the idea of a universal body is not only impossible but a notion that benefits those to whom embodied privileges apply—white, male, heteronormative, First World, affluent (e.g., Conboy, Medina, & Stanbury, 1997). For example, one author recognized the importance of within category variation and differences, rather than pointing only to differences between identity categories, ceased dividing women into those who embodied a hyper feminine style and those who did not, and proposed positing a spectrum of femininity (Dosekun, 2015). Write field notes or write in a research journal and note participants' commonalties *across* gender or other divide. Then note differences in embodied performance of self *within* gender or other group. Use these notes and comparisons to enrich the portrayal of participants as individuals and as members of group(s), reflecting on intersectionality of identity.

Catch a Wave

A main contention of this book is that bodies exist in continual flux, with body-selves always becoming yet never arriving at a static state materially, psychologically/cognitively, or spiritually (see Chapter 1). When we do qualitative research, we capture details of only tiny slices of time, yet we present our findings as though we have faithfully represented who our participants *are* and what they *do*, or at least how they were and what they did during our period of data collection. Yet regardless of the duration of data collection, it is simply not true that participants' bodies, identities, practices, cultures, and contexts remained constant. I, too, have struggled to show dynamic being and change while also authoritatively and persuasively arguing for a coherent set of findings in qualitative research written in research report form, especially since I also do not want to present my participants as having existed in X state when my study began and progressing to Z state when it concluded. One goal of doing embodiment, then, is to attempt to capture (or at least acknowledge) flux as it happens, in moments of transition and change. Try to show processes not just outcomes, change happening not just change accomplished, continual adaptation and evolution. Arguably, narrative and personal reflection are the genres most easily mobilized for dynamic portrayals (Yamasaki, Geist-Martin, & Sharf, 2016). Nordmarken's (2014) autoethnography, for example, shares and reflects on everyday interactions during the author's transition from female to a "more masculine" body via a "gender-ambiguous" body. The transitional time shows gender as liminal, shifting, and evolving. In a more traditional research report format, Draper (2003) investigated men's embodied responses to their female partners' changing bodies and selves as they experienced pregnancy, birth, and breastfeeding. Draper emphasized the process of the partners' bodies growing larger as the pregnancy advanced, the process of labor, and process of a baby emerging to become its own separate person who is fed by the (ever changing) body of the female partner. Finally, Rice's essay (2015) discusses the biopolitics of the current moralistic emphasis on the healthy body as subject only to individual effort and control, ignoring systemic inequalities and alternative conceptions of beauty. Moreover, Rice argues that current biopolitics ignore the fact that bodies are always in flux, and she suggests a model of a "body-becoming" pedagogy that highlights the persistent instability of all embodiments.

Expect the Unexpected

One of the most delicious aspects of qualitative research is the frequency with which I encounter unexpected opportunities of fieldwork, unanticipated tangents in interviews, and unsuspected patterns within analysis. In my experience, queries about bodily experiences and our interpretations of them more often than not spark particularly rich and fascinating responses. Granted, I may be more intrigued than others by vulnerable bodies in particular, given my extensive health care experiences and my embodied empathy for pain and illness. In my favorite classic commentary on the emotional toll of qualitative research, Rothman (1986) explains that she thought she had planned a straightforward set of survey and interview questions with anticipated answers (grounded in extant research) about pregnant women's experiences of fetal genetic testing. With her own infant at home, Rothman found herself viscerally empathetic with women faced with genetic abnormalities and the decisions they made on whether to terminate their (desired) pregnancies; the focus of Rothman's research shifted dramatically in response. In another study, Thorp (2006) found her planned curriculum didn't work, so she surrendered control and found herself in the embodied process of co-creating and caring for a garden on her participants' terms, at a diverse, underresourced primary school that served urban children. Her embodied representation of the project includes photos, drawings, and children's words, alongside the researcher's thoughts and feelings in a deeply moving account. While you can't know what will happen during your own project, you can be open to embracing unexpected outcomes, trusting that bodies will make themselves known in intriguing ways. You can start by journaling expectations you have at the beginning of the project of how embodiment will play out so that you can see what changes later (and what does not).

Start with Your Own Body

Autoethnographic explorations of embodied experiences proliferate (see Holman Jones, Adams, & Ellis, 2016). Even for scholars whose primary focus is on their participants, writing autoethnographic narratives, poetry, or other artistic representations can open the door not only to understanding how the political is embodied and our bodies enact knowledge production through our lived experiences, but also to richer understandings of participants' lives and worlds as they converge and diverge with researchers' body-selves. Even autoethnographic accounts that go unpublished shape researchers' embodied sense making processes and have value as part of the research process; simply rendering one's embodied experiences aesthetically offers another perspective and enriches research outcomes (Ellingson, 2009).

Note

1 Saldaña (2014) also dismissed (with great wit) terminology that proliferates in this very book—"If you use any combination of the words *body, bodies, bodied, bodying, embodied, embodying*, or *embodiment* more than five times in one paragraph, you need to bring it down a notch" (p. 979; original emphasis). I sheepishly admitted to him via email my impending guilt for doing one of the practices he specifically cautioned readers against, and he graciously acknowledged both my praise of his article and my goal of engaging with theorizing of embodiment in order to render it useful for a wide range of qualitative researchers, not only the theory-wielding elite.

4

ETHNOGRAPHIC BODIES

Enacting Embodied Fieldwork

Doing Legwork: There's No Rest for the Wicked

Dr. Armani, head of the Interdisciplinary Oncology Program for Older Adults, loved to tease me, and he and I had several go-arounds about Catholicism before this one. That morning he sauntered into the clinic at 11 a.m., quite late because he had been on rounds in the main hospital. The space, always crowded, felt particularly full that morning, and Dr. Armani's patients were sharing an examination room hallway with Dr. Munson's patients. One of the computers wasn't working, and a young man from the information systems department was trying to fix it, while Annette, an administrative assistant, leaned over him discussing the problem. A cart full of charts partially blocked the hallway, and as the nursing assistants, geriatric oncology team members, and Dr. Munson's primary nurse and nurse practitioner tried to move around, we all kept bumping into each other.

"Good morning, Laura!" boomed Dr. Armani as he entered the busy desk area, mischief dancing in his dark brown eyes.

"Good morning, Dr. Armani," I replied, smiling. I had been up late the night before working on a paper, and I was tired and my right knee ached horribly. But I was determined to be cheerful.

"You know, Laura," he said in his strong Italian accent, and I could see the bait coming. "I was thinking about all this crap you said about sex and gender and communications," he explained gleefully. "And I was thinking that the Catholic church is really the *most* feminist because of its focus on the Virgin Mary."

The former junior varsity collegiate debate champion in me bit – hook, line, and sinker. "*No way*," I said. "The only reason they like her is that she didn't have sex. That's hardly a feminist position."

"But they elevate her above all the saints," said Dr. Armani, warming to the topic. Beth, one of the nurses, walked by, shaking her head in amusement.

"They elevate her because they are *fixated* on her virginity," I exclaimed vehemently, oblivious to the rising volume of my voice. "They blame women for original sin, and see Mary as the only woman who somehow managed to avoid temptation. I hardly think—"

"That's not true!" shouted Dr. Armani, thoroughly enjoying my irritation. He beamed at me, his eyes sparkling. Neither of us was paying any attention to Dr. Munson, whose pale face was now red with anger. "The Virgin is valued as a woman because—"

"Would the two of you stop this!" yelled Dr. Munson. I instantly cringed in humiliation, wishing the floor would open up and swallow me, but Dr. Armani just laughed. This made Dr. Munson even angrier. "You aren't even talking about a patient! And I can't hear anything," he continued, "I can't even dictate!"

"OK, OK," said Dr. Armani soothingly to Dr. Munson. "No problem." After a last glare at my crimson face, Dr. Munson returned to his dictation. Dr. Armani turned to me and continued in a quiet voice, "We'll talk about this more later." He winked at me and patted me on the arm. I could still feel the heat in my face, and my discomfort must have been obvious to Dr. Armani because he put his arm around my shoulders and gave me a squeeze. "Don't worry." Beth approached him with a stack of prescriptions that needed his signature, and I wandered over to the opposite side of the clinic from where Dr. Munson sat, shame still heating my face. Susan offered me a sympathetic glance and patted my shoulder as she walked by on her way to the photocopier. I was horrified at my chastisement but also worried that this might interfere with my access to the clinic.

After that day, I made it a practice to give Dr. Munson as much space as possible. Weeks later, I found out that his mother was dying in the hospital when he exploded at us that day, and he was obviously under a great deal of stress. Although he nodded at me politely after that day, and once even asked kindly about my knee brace, I still figured that keeping my voice down and my body as far away from his as possible was my best strategy.

Bodies in the Field

Ethnography

Ethnography is a broad methodology that encompasses many practices, approaches, and philosophies. The methodology is difficult to define, and no consensus exists within communities of ethnographers on its ideal definition or description (see Atkinson, Coffey, Delamont, Lofland, & Lofland, 2001; Potter, 1996). The method assumes a naturalistic paradigm (Guba & Lincoln, 1985), meaning that it involves studying groups of people in their real-world contexts. Ethnography's central strength is its "potential to reveal the unanticipated loose ends and discontinuities of everyday life [that] are critical to a deeper understanding of social complexity" (Seymour, 2007, p. 1189).

Ethnography requires being present in the (traditionally material and now also virtual) space being studied, for the ability to make claims is grounded in researchers' direct observation of and participation in that space, usually for an extended period of time (e.g., Lindlof & Taylor, 2010). Being "there" and writing about what one sees, hears, feels, smells, tastes, and otherwise senses provides the ethnographer with the basis for analysis. Within a poststructuralist perspective on qualitative methods, "being there" in the ethnographic field is a fuzzy process, fluid, with the emphases on process, participation, and ongoing "becomings" of embodied and emplaced body-selves, including those of researchers. "Fieldwork is itself a 'social setting' inhabited by embodied, emotional, physical selves. Fieldwork helps to shape, challenge, reproduce, maintain, reconstruct and represent our selves and the selves of others" (Coffey, 1999, p. 8). Ethnography incorporates both *emic* or participants' perspectives and *etic*, or outsider perspectives (Potter, 1996). Fieldnotes as data and all representations of ethnographic work—in any genre or medium—should convey "thick description" of the people and culture studied, including rich, specific details of embodied (verbal and nonverbal) communication (Geertz, 1973) and intra-action of people, actants, and discourses (Barad, 2007).

Ethnographer Bodies are Research Instruments

The ethnographer as research instrument is a truism in qualitative methodology; ethnographers take in details of our participants and setting using our bodies' senses as filters and our bodies as permeable containers. Ethnography contrasts with other types of instruments covered in research methods courses, such as surveys, experiments, and tests of biometric indicators (e.g., stress hormones in saliva). Yet researchers often conceptualize what ethnographers do primarily as a mental exercise of discerning among the many sensory inputs in our field setting to determine which are worth attending to, whether briefly or for an extended time, at any given moment. In a busy medical office setting, for example, an interaction between colleagues, the technology used in professional practice (e.g. a stethoscope to listen to heart beat or breathing), or the way participants move within a confined, crowded space would all warrant attention but could not all be attended to simultaneously. Attention leads us to compose fieldnotes about what we see and hear, and ideally also about what we smell, touch, taste, and experience more holistically. Researchers can expand our understanding of ethnography by considering how we do embodiment, our participants do embodiment, and our modes of doing intra-act with one another and the objects and place(s) of field settings. The focus of ethnography may be a topic that explicitly relates to embodiment, such as Hopwood (2013), who studied embodied work of childcare and parenting in a context of a family resource center, or embodiment may be more of an ethnographic practice issue that is noted in fieldnotes and reflections on the ethnographer's presence and the sense making processes, such as my own unruly body to which I attend in all ethnographic settings.

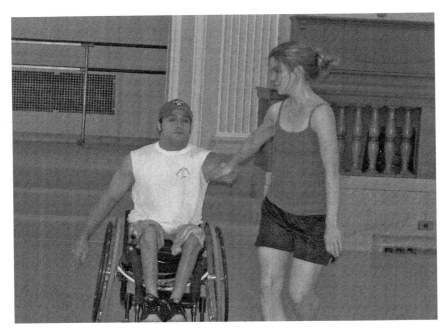

FIGURE 4.1 Dancing Wheels.
Ethnographer Margaret M. Quinlan dances with a participant during her study of a professional dance company, Dancing Wheels, that integrates stand-up and sit-down (wheelchair) dancers.

Courtesy of Mark Daurelio and Margaret M. Quinlan (participants), and Mary Verdi-Fletcher (photographer).

Ethnographers' bodies/instruments intra-act with participants' bodies in specific places. Ethnography can thus be understood as "a bodily and material 'conversation' with the field, with the other . . . [and] fieldwork practice [conceived] in terms of bodily doings and sayings amid, and attuned to, material environments" (Hopwood, 2013, pp. 228–229). Such explicit framing of ethnography within material being and place enables researchers to make explicit the embodied aspects of our sense making, regardless of research topic. Sensory ethnographers urge researchers to wade into the mess of everyday life and explore how people and objects co-exist and mutually constitute within the material world (Pink, 2009). For example, a sensory ethnographer of cooking describes

> a sensual perspective—centering, decoding, reframing, discovering, and discoursing the clutter of the Made World, literally as "embodied" participants and observers, full of touch, smell, taste, hearing, and vision, open to the buzz and the joy and the sweat and the tears—the erotics—of daily life.
> *(Brady, 2004, p. 628)*

Of course, the smells and textures and bodily movements are present in any ethnographic site, so the choice is to attend to them, shifting one's attention (partly) away from discourse (which admittedly may be particularly pronounced in my work due to my training in the disciplines of communication studies and gender studies) to include more emphasis on materiality.

Embodied fieldwork not only resists the mind/body split, but also takes seriously emplacement of the ethnographic site, or the ways in which participants' behavior is contextualized within specific types of spaces, configured with various technologies and other objects, in which the bodies of the ethnographer and participants intra-act with actants (Shilling, 2012). For example, an ethnography of fashion clothing stores examines the feminized retail sales associate occupation, including the emotional labor required and the gendered occupational segregation that leads to women comprising the vast majority of these workers. In addition, Pettinger (2005) positions the retail objects (i.e., clothing) as crucial to meaning in this setting because the clothing is highly gendered (and marked by age and class as well) within the sticky web of culture in which it is designed, manufactured, and marketed. Thus the femininity (or masculinity or androgyny) of the actants (i.e., the clothes, the store decor) participate in the ongoing gendering of retail shopping spaces that are also work places (for participants), leisure/recreation spaces (for customers), and sites of commerce (for both). Moreover, "[b]odies are always located" (Longhurst, Ho, & Johnston, 2008, p. 208), and sensory experiences are always "emplaced" in a cultural, historical, and political milieu, with dynamic history (Pink, 2009). In Pettinger's case, the politics of gendered work and gendered fashion manifest within specific retail settings, and also in the patriarchal history of sex and gender that contributes to and results from gender politics, that is, gendered expectations for emotional labor in service sector jobs, gendered employment patterns within service positions, and the gendering of retail products such as clothes (Pettinger, 2005).

"Participant observation" is one of the terms used to describe ethnographic fieldwork, emphasizing the participatory nature of the researcher's role in the context being studied (Lindlof & Taylor, 2010). Awareness of the ethnographer's participation in the field setting is key to situating oneself in an ethnographic field site and understanding that the ethnographer and participants are constituting each other, the place, and the meanings of their shared tasks or purpose(s). Sensory ethnography emphasizes

> researchers' empathetic engagement with the practices and places that are important to the people participating in the research. And by association it does *not* therefore principally involve the collection of data *about them* that can later be analyzed. Rather it involves the production of meaning *in participation with them* through a shared activity in a shared place.
>
> *(Pink, 2011, p. 271; original emphasis)*

A great example of a researcher's awareness of and keen insight into her own embodied participation in making meaning with her participants is Nabhan-Warren's (2011) fieldwork in a Christian faith group.

> I listened to what my body wanted to do—to dance (the music was intoxicating)—and in moving around the backyard, arms linked with other faith course candidates, I gained an insight into my interlocutors' beliefs and desires that would have been impossible were I to have sat on the sidelines, observing and taking notes. I experienced their happiness, and the emotional release they felt as they bobbed and weaved through the backyard space. . . . After dancing, we sat, sweaty and tired, and sipped from bottles of water.
>
> *(Nabhan-Warren, 2011, p. 392)*

In this excerpt, the ethnographer's body has agency and desire; her body-self wants to dance. She describes emotions in her body, as well as a physical urge to drink water and an awareness of her body's exertion (sweat and fatigue).

A multisensorial approach to ethnography enables fieldwork to be "neither dominated by not reducible to" vision (Pink, 2009, p. 64), as Nabhan-Warren's text makes clear. The tendency of ethnographers to privilege sight as a sensory capacity is perhaps unsurprising, given the framing of ethnography within visual terminology (i.e., observation). The fetishizing of vision has been much maligned by some theorists as a Western tendency, certainly of those socialized in and through the English language. Seeing is equated with knowing and understanding, e.g., I can *see* what you mean; I can *see* that the data do not support such a conclusion; can you *see* his point of *view*? Western cultures reinforce a *primacy of vision* (particularly as seeing done by a satellite or other disembodied technology) as an epistemological stance (Haraway, 1991) and *ocularcentrism* in our sense making (Jay, 1994). Yet to see (or any other singular sense) is not a simple or straightforward perception as it has been imagined through the lenses of positivist science. Instead, drawing on Merleau-Ponty (1962), Thoresen and Öhlén (2015) suggest that "observing is not about seeing the world as it is; observing is reciprocal between what is seen or observed, and the observer. Knowledge is developed through the lived body, not only through the eyes and the mind" (Thoresen & Öhlén, 2015, p. 3). So vision is not only one of many senses through which ethnographers know (and are known by) the world, but seeing itself is grounded in a complex bodily sense making process that involves not just the eyes, nerves, and brain but the whole body-self.

Embodied, sensory ethnographic practice starts with simply taking up space with our bodies; our physical presence must be negotiated with the gatekeeper(s) of our field sites, which may not be a simple feat.

[T]he ethnographer has to sit or stand or lie or be *somewhere*. . . . A space has to be made, or found, or negotiated for the body-thereness of the ethnographer. . . . There may be no proper, ready-made or appropriate place for the body to co-locate. . . . It is often difficult to know what to physically *do* in the field in order to look natural, comfortable, engaged and welcoming, while not appearing bored, threatening or judgmental. Immersion and integration are physical aspects of fieldwork.

(Coffey, 1999, p. 73; original emphasis)

Hopwood (2013) reflected on the difficulty of knowing where to put his large, adult, male body in a pre-school, where his body was so much larger than the kids' bodies. He did not want to stand too close or too far away, and he struggled to fit his body on the kid-sized furniture and in small spaces.

Ethnographers should not only document others' sensory experiences but learn to recognize opportunities to develop understandings of participants through multiple senses (Pink, 2009). For example, one researcher sat in a hallway at a residential hospice for an extended time, unable to see into individual rooms. She discerned that lived experience of spaces in fieldwork give us insights into not just what we see but what we *cannot* see.

[W]hat I could not see, but could hear, smell, and sense, had to be the same for many of the patients. This is because most patients spent nearly all their time inside their rooms, but often with the door left open. This meant that contact with the outside was through listening to, and to make meaning of, the sounds of people, machines, or unidentified noise.

(Thoresen & Öhlén, 2015, p. 1592)

Furthermore, the roles and relationships ethnographers form during fieldwork are tied to our bodies and senses (Coffey, 1999). Nonverbal communication—appearance and grooming, body language (e.g., gestures, posture), facial expressions—arguably "is *the* most important component of behaviour in public" (Shilling, 2012, p. 85; original emphasis) because "[t]he ability to 'look' as we are expected to (or not!) is a key factor in our ability to conduct the research; to promote trust and reciprocity" (Coffey, 1999, p. 71). In this sense, "the ethnographer's body can be a barrier, but also a bridge, to cultural exchange and to entering and knowing her interlocutors' lifeworlds" (Nabhan-Warren, 2011, p. 380). For example, Turner and Norwood (2013) discussed their researcher body-selves as constituting a variety of roles in helping (and hindering) their efforts to establish relationships with participants. Their ethnographic bodies serve as a (presumed) impetus for studying topics that related to their own gendered embodiment and own birthing experiences, an instrument of research when bodily knowledge became part of the research conversation and the researcher body a point of interest and discussion, and an impediment to research when the researchers' bodies were understood to be critiquing the bodies of participants on gendered topics of birthing, breastfeeding, and transgender identities.

Moreover, ethnographers sometimes have to figure out how to do our bodies in quite specific, stylized ways that are compatible with the bodily performances of our participants and that foster our relationships; ethnographers "'learn' the craft skills of body work during fieldwork participation" (Coffey, 1999, p. 73). This could involve learning social practices, adapting ways of speaking and relating to fit cultural norms surrounding, for example, (differences in) gender or age. Violating social norms could jeopardize a researcher's role and relationships with participants when greeting another with a hug, kiss(es), handshake, bow, and salutation appropriate to the person's social standing, especially in places such as Japan where greetings acknowledge gender, age, and status (Kondo, 1990). Embodied relationships between researcher and participants also may reveal our (and participants') humanity as we interact with others through witnessing and deeply sharing each other's embodied being in a given moment.

> [B]y taking seriously the body as the locus of experience the researcher engages composite modes of being, including emotional comportments, expressions, postures, movements and touch. In this way, intriguing and sometimes challenging relationalities might emerge between the researcher and the researched. Furthermore, it is through these embodied interactions that moments of mutual vulnerability might surface, where precariousness unfolds, and where distinct lifeworlds may briefly interlock.
>
> *(Hoel, 2013, p. 46)*

While such intense moments of connection certainly do not happen routinely, ongoing relationships can foster deep engagement and receptivity so that when they do occur, ethnographers can appreciate special moments (data "hot spots," MacLure, 2013) and the insights and opportunities they generate.

Embodied Emotions in Fieldwork

Sensing and communicating emotions forms an integral component of ethnographer bodies as research instruments. Emotional expression is integral to the establishment, maintenance, and ongoing development of relationships with participants. Emotions are

> embodied biocultural processes between and within persons. These both act on and shape cultural meaning systems that support the person in appraising her or his own and others' experiences. They assist social actors to adequately communicate and navigate their social and cultural environments.
>
> *(Stodulka, 2015, p. 85)*

Embodied emotions enmesh ethnographer and participants "not as bodies with clearly delineated borders, but as fluid bodies, as bodies of blood and water, relating to other bodies situated in complex social contexts" (Hoel, 2013, p. 39). Ethnographers learn how to perceive, interpret, and express emotions, which "are not substances to be discovered in our blood but social practices organized by stories that we both enact and tell" (Rosaldo, 1984, p. 143), as we interact with participants. The ethnographer's emotions not only evoke and respond to those of participants but "shape the ways in which stories are told and social realities are conveyed" (Stodulka, 2015, p. 86), both in fieldwork interactions and in data construction (writing fieldnotes, reflections, and journal entries), analysis (coding, memo construction), and representation (narratives, reports, performances), although all but trace evidence of the impact of emotions on findings may be removed in traditional and even interpretive research reports or critical analyses so as to meet postpositivist standards of detached prose.

Emotions have been disparaged under (post)positivism. The rational/emotional dichotomy is a falsehood perpetuated within the academy and Western societies more broadly; the belief that logic is or can be somehow completely or even mostly detachable from our emotions has been disproven by neuroscientific studies (Strean, 2011). Researchers in the field may suppress emotions at times, but emotions are never not present. Rationality is praised as a good way to process ideas, and so researchers have often emphasized the rational to the marginalization of emotional experience (Richardson, 2000).

Experience is not only always already emotional, but also deeply embodied in its expression. When we are sad, tears leak out of our eyes, some of our facial muscles contract, and we make sobbing sounds with our lungs, throat, and larynx. When we are excited, our hearts race, our torsos lean forward, and our eyes focus intently. When we are angry, our shoulders tense, the muscles on our foreheads contract, our hands may shake, our voices may get louder or eerily soft. Expression of certain emotions may or may not be expected, welcomed, or appreciated within our ethnographic sites, but they will certainly be felt, often deeply, by ourselves and our participants. Ethnographers experience emotion bodily as part of our sense making processes, both of our own and of others' experience (and the intersection of them). Ethnographers can choose (to some degree) how to acknowledge their emotions in fieldwork; we can resist the imperative to focus on our seemingly rational moments and instead understand ourselves "not as autonomous, rational academics, but as people who sometimes experience irrational emotions including during the course of research. Emotions matter" (Longhurst, Ho, & Johnston, 2008, p. 213).

Moments in fieldwork allow researchers to examine "how private and social experience are fused in felt emotions" (Ellis, 1991, p. 23). Stodulka (2015), for example, conducted an ethnography with a group of male street youths in Yogyakarta, Central Java. Reflecting on his own emotional responses to being emotionally manipulated and pressured by some of his participants generated crucial insights into these young men's strategies for survival on the streets through an "emotional economy" through which they "transform[ed] scarce resources, marginality, and stigma into emotional bonds and vital socioeconomic co-operation with various actors of their widespread social networks" (Stodulka, 2015, p. 91). His own emotions of disillusionment, frustration, disappointment, and anger manifest deeply within his body and helped him understand how the young men's emotional exchanges worked. The youths "acquired and continuously refined their social skills of empathically assessing and framing encounters with various interaction partners" (p. 2) for their own gain. Stodulka realized that he too had used his emotional bonds with these young men for his own gain, primarily to learn their cultural experiences in order to generate journal articles and academic employment, and that other researchers, volunteers, and NGO workers also engaged in an emotional bartering process with these youths, pursing their own scholarly or professional service goals *and* seeking to help the youths at the same time. Stodulka suggests that it was his embodied emotional insights that led to this understanding, rather than his more rational notes on the youths' relationships, language, and behavior. The emotional exchanges happened within the context of Javanese culture, particularly that of impoverished urban communities. In the next section, I discuss the role of culture and social structures in understanding ethnographic embodiment.

Cultural Context and Social Structures

Local experiences of individual bodies are understood within emplaced interactions, but also within global discourses of power, privilege, oppression, and resistance (Ritenburg, Young Leon, Linds, Nadeau, Goulet, Kovach, & Marshall, 2014). Of course, ethnographers generally cannot point to a specific moment in which, for example, a boy internalizes messages about heteronormative masculinity and adopts aspects of the prevailing masculine ideals to his own bodily performance; yet we know that he schools his expressions and comportment to conform to (and perhaps resist) dominant messages about masculinity that circulate in and through specific bodies in his day-to-day life (Kimmel, 2004). Ethnographers document what our participants do but without claiming direct causality, that is, that the culture made him do it—e.g., he sits with his legs wide open, freely taking up significant space in a crowded subway car at rush hour, simply because society awarded him masculine privilege—diminishing the agency of participants and ignoring the complexity of personality, agency, and constraint. At the same time, ethnographers should not ignore the influence that circulating norms have on shaping perceptions of appropriate choices for embodied performances of self, gender being but one example. How can ethnographers bring together the individual, group, and organizational experiences of our embodied participants with scrutiny of cultural norms, practices, and influences, thereby attending to both the material and the discursive? How do we document grand narratives operating within local spheres when they are contextual and interwoven with embodied experiences rather than causal? Possibilities for examining how culture circulates in and through bodies are a primary concern for critical approaches to studying embodiment, particularly for decolonizing methodologies (Ritenburg et al., 2014). Several insights about the relationship between individuals and cultural structures and contexts may be helpful for researchers doing embodiment in ethnographic research.

First, no embodied experience is without both power and resistance; individuals have neither completely free choice nor an utter lack of agency. Foucault's concept of disciplinary power is helpful in understanding this relationship. He argued that in modern societies, we have moved away from traditional, sovereign power structures in which an authority figure exercised control over subjects through force, regulation, and punishment (Foucault, 1977). Now we internalize disciplinary power and learn to regulate and discipline ourselves. In daily life, we make choices in our clothes, accessories, hair styles, and grooming. In a larger sense, we choose careers, what type of home we have, and life partners. All of these choices are constrained by discourses on what is appropriate in our cultures, for our gender, class, occupation, and so on. We regulate ourselves, and we also practice resistance to regulation. Ultimately, it helps to remember that power and resistance are fundamentally inseparable. One way to represent this inherent connection is as "resistance^power," signifying that the "phenomena exist and often give impetus to each other" (Clair, 1998, p. xv). If ethnographers agree that power and resistance are inseparable, then when we conduct ethnography, we can simultaneously seek signs of both in ourselves, our participants, their places, and the objects with which they intra-act, not totally subject to nor free of discourse but always resisting and always constrained by discursive power. Documenting the tension between resistance and power can be fruitful. At the same time, there are material limitations—income, education, body type, race, gender, (dis)ability—intersecting with these discourses. For example Roberts (2013) observed "black and brown bodies" of hip hop dancers and noted systemic racism, classism, and pathologization of bodies as sites of deficit and disorder. She posits:

> To be oppressed is more than a feeling. The physical sensation of weight bearing down on bodies, and the sharp, staccato movements, postures, and gestures that are produced lead to nuanced understandings of how knowledges of macro level-structural injustices live and are animated in and through bodies. I am pointed in the direction of a collective reality, rather than an individual experience.
>
> *(Roberts, 2013, p. 285)*

The embodied links among racialized bodies, particular movements, and macro-level injustices are rendered into an aesthetic experience through dance-as-storytelling and dance-as-self-making.

Another productive way of understanding the material and individual experience is acknowledging the dynamic nature of body-selves in the field. We have many aspects of self that we perform in different ways to varying audiences and constituencies in our lives (Goffman, 1959). As ethnographers, we do well to remember that our participants have just as many aspects of self, and that they are not static but fluid. To illustrate, I share one ethnographer's acknowledgment that her participants

> occupied and moved between *several* subject positions such as: intimate lover, wife, second wife, mother, divorcee, widow, recovering drug addict, and victim of marital abuse or rape. I spoke to abled bodies and disabled bodies, terminally ill bodies and healthy bodies, broken bodies and healed bodies, women who self-identified as feminists and anti-feminists, gender-activists and ANC[political party]-supporters, Malay, Indian and coloured Muslims, and so on. This subjective elasticity is illustrative of the messiness of embodied and lived realities that are continuously produced and in process.
>
> *(Hoel, 2013, p. 33; original emphasis)*

Attending to the fluid manner in which participants embody multiple, intersecting identities avoids essentializing participants within a single identity. Manifestations of culture become evident as individuals move between their performances and interact with others whose performances are equally multiple and shifting.

> Furthermore, relevant social structures are continually in flux, never stable.
>
> Structures and contexts are not . . . remote, monolithic entities, somehow removed from everyday social life. They act in this social sphere to be sure, but they are also acted upon in ways that leave them open to alteration or outright removal. It follows that the social and structural forces . . . must themselves be established, supported, made and remade in time and space. . . . social contexts are assembled and reassembled in the activity of diverse bodies, spaces, processes, and objects.
>
> *(Duff, 2012, p. 284)*

From a communication perspective, structures are less fixed entities (organizations, laws) and more continual processes of (re)organizing (Miller, 2014).

Additionally, social structures are imbued with discourses of identity. In ethnography, identity factors may fade into the background of ethnographic fieldwork and suddenly come to the forefront in moments that implicate the materiality and relevance of say, gender or race, only to fade again. Holmes (2014), for instance, observed a little girl on a school playground, running freely at recess only to be captured by a group of boys who grabbed her by the arms and restrained her. Gendered discourses are always circulating, always being disciplined by us on ourselves and thus supporting their influence on others; they are never absent or irrelevant, no matter how much some people may want to believe in a world that is post-gender or post-race, ourselves and our spaces are heavily regulated by these identity discourses (for good and for bad). While our research purposes generally do not implicate such background identities directly, they are always present and periodically attending to them is critical to understanding participants' worlds.

Practices for Embodying the Field

Mark Unmarked Bodies

Unlike in earlier eras, more culturally powerful people have the privilege of remaining *un*marked by difference. Throughout the nineteenth and much of the twentieth century, the primary way to indicate wealth and power was to have a robust body (while others went hungry). Now it is the ability to fit within the realm of the normative (i.e., presumably typical/common *and* culturally preferred)—as male, heterosexual, cisgender (i.e., conform to cultural norms for masculinity or femininity that correspond to one's biological sex), white, middle or upper class, nondisabled, a native-born citizen—that denotes cultural privilege (Thomson, 1997). It is easy for identity and practice markers to go unmarked by ethnographers when normative categories fade into the taken-for-granted elements of a culture, such as noting race only for racial minorities, while leaving whiteness uninscribed. Likewise, cross-sex (heteronormative) demonstrations of affection, such as holding hands in public, may be described without labels such as straight or heterosexual, whereas same-sex interactions are likely to be called out as indicative of a lesbian or gay relationship. This is not to say that ethnographers should stop noting those with minority characteristics, but that we should bring to the surface unacknowledged aspects of bodies—such as their ability to move in certain ways, or their ability to pass unnoticed in some contexts, or their assumed normalcy of size or comportment. Challenge yourself as you engage with participants to describe the tones and textures of whiteness, portray the gait of nondisabled bodies walking, consider what aspects of appearance or manner lead you to assume that someone has middle-class economic status. By marking the powerful and purportedly normal, ethnographers frame differences as inherently relational; that is, difference is a comparison between two or more individuals—one person is Latina and one is Pacific Islander—rather than locating the difference *within* the member of the less powerful group—that person is poor (and therefore different, because "normal" people are middle-class) (Minow, 1990).

Openness to the Unplanned

While ethnographers absolutely can and should make guiding goals and principles of our research, we also should be open to the unplanned and unexpected, because these often yield critical insights. One of the strengths of ethnography is its flexibility (Seymour, 2007). As researchers and participants build relationships, opportunities for research can shift and expand, from a researcher's interest and from the changing needs of participants. More than a few academics are control freaks—myself absolutely included—and so we may feel uncomfortable deviating from our plans for conducting fieldwork, such as shadowing specific participants for a certain number of days each or conducting interviews at a particular point in the fieldwork. Yet the willingness to go with the flow of action, to appreciate the flux of people, place, actants, and researcher's body-self, can pay off with additional insights into practices or ways of being among participants. Incidents, emergencies, mundane practices you had not witnessed previously and visitors to a setting, for example, can spark interest and opportunity in a particular area. Unplanned moments may center on bodily failures—failing performances open up opportunities for understanding the typical performance of self (Goffman, 1959). For example, I have often felt self-conscious, even shame, about my inability to stand for long periods of time due to my impaired, and now amputated, leg during fieldwork, especially when not wanting to call attention to myself. Yet the need to sit down led me to sit at the nurse's station as a break from roaming the dialysis clinic during my ethnography of communication among staff and patients in that place. Sitting at the nurse's station gave me another perspective on the surrounding action, as did the staff break room to which I retreated to drink water, use the restroom, or just rest briefly. In these places I conversed with staff as they completed tasks other than direct patient care, enabling me to learn more about their responsibilities, including to log information on the computer at the nurse's station, and their banter with one another when "off the floor," i.e., in nontreatment areas where patients could not see or hear them (Ellingson, 2011a). Serendipity is part of the ethnographic research process, temporally, relationally, and analytically, and ethnographers can capitalize on opportunities by being prepared to welcome the unexpected (Fine & Deegan, 1996).

Illustrating Actants

Make a concerted effort during fieldwork to take photos of or sketch actants (nonhuman objects) in the places you are studying in order to visually record them. If possible, touch, smell, lift, and otherwise interface with the objects in order to describe them in as much sensory detail as possible—temperature, texture, heft, scent, vibrations, color, utility—enriching your potential for analysis of how they intra-act (Barad, 2007) with people, cultural discourses, and other aspects of the scene. In Birk's (2013) autoethnographic account of chronic pain, for example, pain medications function as powerful actants whose intended effect (i.e., diminishing pain) and unintended effects (e.g., fatigue) greatly influence the range of Birk's activities and understanding of herself in ways that are interwoven with her bodily sensations and responses. Thus describing the color, shape, texture, and size of the pills, their dosage, container, and so on are crucial sensory details of the actants.

In a very different ethnography, global poker [card game that involves gambling] provides an exemplar of how a large number of actants—the Internet, mobile devices, and various telecommunications technologies—create the very setting(s) that become available (and constantly shifting) settings for ethnography. Poker

> is spectacular, too, in its sheer diversity of media forms, from syndicated television shows, online poker sites, world poker tours and regional tournaments to numerous mobile, digital and software devices. . . . Whether playing or watching, poker enthusiasts have access to an immense spectrum of media . . . The interaction of these technologies and their human participants constantly changes how the game is reported, played or watched.
>
> (Farnsworth & Austrin, 2010, p. 1121)

In this new media ethnography, actants intra-act in fascinating ways with people, places (particularly the international gambling hub of Las Vegas), and other technologies (e.g., televisions). Farnsworth and Austrin focus their attention on describing the capacities of actants—technologies, their user interfaces, and ways of talking online and in person about play, spectatorship, and gambling as they are carried out in virtual and physical spaces.

Create a Model

Janesick (2000) recommended that researchers construct a model of "what occurred" in their research site

> because developing a model comes close to choreographic work or artistic work and serves as a heuristic tool. . . . like the scene designer or architect who builds a model, the choreographer or dancer who captures the dance on film, or the artist who creates a drawing or series of drawings, the researcher can use the model as a tool for further work or it can serve as a simple historical record.
>
> (Janesick, 2000, p. 386)

Such a conceptual model offers a great way to see your participants' world holistically. Try to illuminate participants' relationships, practices, and processes spatially on paper, using PowerPoint tools, or using software designed for making graphs, models, and charts (e.g., Smart Draw). Even if this representation does not end up in publishable form or constituting specific findings, it will influence your holistic understanding of your ethnographic project and may generate vital insights. The act of shifting to and reaching for a different tool and medium can really shake up unexamined assumptions and expectations of your participants and their world. Nursing scholarship provides a lot of examples of models. Tanner's (2006, p. 208) model of clinical judgment in nursing features a figure reflecting different phases and process in nurses' decision making. Likewise, Leininger (2002) details her method of "ethnonursing" as a way to study nursing practice and culture and provides a model of her Culture of Care theory, which visually represents complex relationships and processes. Both of these models use shapes and graphic design to illustrate the textual arguments made by the authors. Ethnographers can fashion much simpler or cruder models as stepping stones toward understanding the complexities of their participants' worlds.

Walking Sounds Fieldwork

Attending to sound in fieldwork includes so much more than just human speech (Paterson, 2009; Pink, 2009). Fieldwork should involve focused attention on the "noisiness of the everyday" (Hall et al., 2008, pp. 1019–1120). This can be accomplished through the technique of "soundwalks [which] are about reactivating one's sense of hearing by attending to the sounds of the everyday, foregrounding a background context to daily life" (Hall et al., 2008, p. 1120). Such ambient noise and background noises are felt not just in your ears but as vibrations throughout your body-self. Moreover, what we hear interweaves with (or stands out as discordant to) what we see and touch and how we move, influencing how we feel in a particular place. Audio ethnographies and creative sonic fieldwork projects include soundscape recordings, soundwalks and sonic maps, radio diaries, and audio essays (Makagon & Neumann, 2009). Makagon and Neumann's website, recordingculture.org, has links to excellent examples of audio documentaries. The book contains great suggestions for getting started, including what types of equipment to use and how to collect audio recordings for your (and your students') research projects. For example, Gallagher (2011) attended to the soundscape of a primary school, noting the ways in which sound—especially demands for silence—reflected the authoritarian use of surveillance, power, and control by the institution, as well as "the disparity between the ideal model of disciplinary power and its everyday functioning in the school" (Gallagher, 2011, p. 52). Soundscapes can also be used as representations of (preliminary) findings; for example, creating "polyphonic sound montages" of interviews of patients at a cardiology department facilitated "co-analysis" as a way to incorporate user perspectives (health care providers and clinic staff) in a study to improve health care delivery (Arnfred, 2015). The sounds of giving and receiving care and of clinic administration offer powerful glimpses of patients' voices and perspectives, enabling listeners to hear individual patients, indicating (perhaps) gender, advanced age, and other elements of particular people's unique voices, which contrast as sound recordings that are far richer than disembodied excerpts of transcripts. Attending to the complexities of background and foreground sounds could be integral to your ethnographic study or could constitute part of your holistic process of participating in and gathering data on a place.

Ethnographic "Go-Along"

Ethnographic "go-alongs" are a combination of informal interviewing and participant observation that involve accompanying research participants as they move through the normal activities of their day, asking questions and observing processes as they happen (Kusenbach, 2003). Such go-alongs contribute rich sensory information that may be not elicited in other circumstances. In Kusenbach's study of urban neighborhoods, she found that go-alongs provided insights into participants' perceptions of their interactions with their environment and community, the ways in which personal biographies or identities were tied to specific places in their daily life, and the complex web of relationships in local communities. Going along is much like shadowing in ethnography, where ethnographers negotiate their degree of visibility as they follow people (often employees of a large organization) to detail their movements and tasks throughout the day. Go-alongs with one participant at a time enable a perspective on activities in the fieldwork setting that differs from one fostered through interaction with people as they come and go in a common space (such as the nurse's station in a clinic or a room of cubicles in an office), complementing a perspective on the group with the rhythms of individuals' movements, and may involve access to different places not yet experienced by the ethnographer's body-self.

Interrogate Silences

As widely advocated by feminist ethnographers, ethnographers can pay attention to embodied silences as much as to what is said because silence can be a form of resistance as a well as oppression (e.g., Edley, 2000; Olesen, 2011). Silence has a variety of functions, purposes, and meanings; it is not merely an absence (Kenny, 2011). Silences are also not merely the absence of speech but embodied practices; "bodies are vessels of meaning and meaning-making, at times, empowered and loud, but also, bereft and silent" (Hoel, 2013, p. 37). A fundamental truth in the communication studies discipline is that you cannot *not* communicate; people always communicate some sense of affect, even if it is complacency, or other mild states. Silence can be powerful and interesting, not just something missing but an embodied communicative practice and meaningful way of being with others (Covarrubias, 2007). Silence can also indicate both strength and physical and emotional pain. Consciously attend not only to the overt activity of a place but to the spaces in between that help give those activities meaning. When do participants enact silence? How do actants producing sounds intra-act with silent people and other objects (Barad, 2007)? For example, ringing phones might move participants to speech. Likewise, internal states might lead to sounds—a participant who is frustrated might emphatically sigh or repeatedly tap a pen against a desk. When participants speak, how and when do they pause and what do those silences do? Spend an entire visit to the field describing the silences. Afterward, write questions for your participants to follow up in (formal or informal) interviews that arise from what you have noted to probe participants for their understandings of silence—as a strategic/purposeful speech act, as a sign of oppression, and as the ambient/unintentional part of a context (perhaps a by-product of another practice). Be sure to note how participants' whole bodies perform silence, not just their (lack of) voice. Note what nonverbal communication cues enact silence—frowning mouth? Intent stare? Blank look? Relaxed face and shoulders? Tightly crossed arms? Frightened, cringing face? Steady eye contact or refusal to make eye contact? Silent tears? Each of these—and endless combinations of nonverbal cues—communicate very different embodied tones of silence into which ethnographers may inquire.

Attend to "Enfleshed Knowledge"

All ethnographers receive messages from parts of our bodies reflecting "enfleshed knowledge," (Spry, 2001), what essayist, poet, and disability activist Nancy Mairs describes as "what I know in my bones" (Mairs, 1997, p. 305), also commonly expressed (in English) as knowing in the gut or the heart. In ethnographic fieldwork, such embodied knowing functions as cues to pay attention. One educational ethnographer described observing children playing and describe the embodied memory of a previous playground incident.

> In the playground on that day, watching this group of excited children, I recall a frisson caused by uncomfortable feelings in the pit of my stomach, tingling and numbness in my arms, sweating, a heavy sensation in my legs. This playground encounter has never been far from my mind ever since. The data enter my body. It seeps in through my skin, my pores, my mouth, my lungs, my muscles, my stomach, my nose, and my fingertips.
>
> (Holmes, 2014, p. 784)

For Holmes, her embodied reaction to a past incident forms a visceral part of the body-self through which she now observes children on another day. The data carried in her body is not a metaphor but a description of how her muscles, stomach, limbs, and other parts respond to what she witnessed and continues to witness, making data (and analysis) something within her rather than an exclusively cognitive process. In fieldwork, such knowing is a prompt that something in the ethnographic site is significant—not necessarily negatively, however; frissons can be prompted by excitement and joy, as well as fear, anger, or dismay. When prompted, engage such a data "hot spot" (MacLure, 2013), make copious notes, and particularly describe your own bodily sensations, trying to capture the quality of your muscle tension, pain, heart rate, and other significant embodied indicators.

Sensory Self-Study

Examine your own body reflexively, taking note of your taken-for-granted ways of moving through the world, including gestures, posture, sitting, standing, walking, and movements involved in everyday practices. Further, ethnographers can become aware of, reflexively consider, and write about the porous nature of the boundary between bodies and their environments, in a process that "is akin to an honest form of Cartesian introspection which identifies and clarifies complex combinations of somatic sensations" (Paterson, 2009, p. 784). The real problem here is the assumption that there are identifiable sensations "out there" (or conversely, felt "in here") reportable in this way (Paterson, 2009, p. 784). Paterson advocates that researchers describe sensory experience in as nuanced a way as possible, resisting that assumption. Acknowledging that interiority and exteriority are a fiction, ethnographers can nonetheless use the conventions of figurative and evocative language to describe (internal) bodily experience by reflecting cognitively on sensations, perceptions, and reactions of bodily parts and systems as though parts and systems could be viewed by the mind and separated from their environment. Such an exercise provides further data and analysis and also assists in the development of a heightened state of awareness of your sensorium as an ethnographer.

Resensualizing Ourselves

Ethnographers can embrace "an ongoing process of resensualizing ourselves" in fieldwork with others by learning new ways of sensing specific aspects of particular places (Brady, 2004, p. 631). The analogy of an apprentice is a commonly referenced role for ethnographers as they learn from their participants about particular sensory aspects of participants' world (e.g., Pink, 2011). An apprenticeship model can be adapted especially to senses; ethnographers learn skills in how to see, how to taste (wine, foods of a variety of ethnicities), to feel textures of fabrics and leathers, to hear different types of cries from babies expressing hunger, fatigue, or frustration. Ethnographers can learn how to listen through a stethoscope in a hospital (Rice, 2010) or how to feel for the best way to access a particular patient's fistula to administer dialysis (Ellingson, 2008). Ethnographers can apprentice to participants to learn how to cultivate sensory skills and then how to understand the experiences that we learn to have, within the cultural language, norms, practices of the group we are learning from/about (Pink, 2009).

How we see or feel with our hands or otherwise use our senses, should be open and expansive not just to, learn new things about the participants' worlds in our usual bodily manner but to learn new practices and capacities for sensing the world. For example, as I write this, I sit in my favorite coffee shop, smelling the roasting coffee beans in the huge roasting machine at the other end of the large open room. The staff know how to gauge appearance, aroma, and temperature of the beans to know when they are done, and I could learn how to scent the aroma of well roasted beans by talking to baristas about how they learned to scent roasting coffee and then practicing it myself.

Write Your Body's Biography

Experienced ethnographers readily acknowledge that we cannot really "separate fieldwork from our own sense of self. In the course of fieldwork we are often engaged in the (re)construction of our own biography and the biographies of others" (Coffey, 1999, p. 68). Remaining focused on your body-self's perceptions and sensations, construct a biography for yourself by describing who you are when in your field site. That is, who are you to your participants? How do you function as part of that particular place? What parts of your body-self seem most relevant in your fieldwork? What has your body learned to do in this place? How has your sense of who you are changed over the course of fieldwork? How do you intra-act with actants during fieldwork? How do you sense discourses circulating through your body-self and the body-selves of participants? This form of reflection may be generative as you connect your evolving body-self with your insights on participants.

5

INTERVIEWING BODIES

Co-Constructing Meaning through Embodied Talk

Doing Legwork: Your/My/Our Bodies

I slowly made my way through a neighborhood of large lots with dense tree cover, navigating around a majestic redwood seemingly plopped in the middle of the road, around which the bumpy path meandered. I pulled up and parked along the road. It was a sunny morning, the beginning to a lovely day. Sharon graciously invited me into her lovely, sprawling house, her happy Bernese Mountain dog, Cotton, eagerly awaiting my attentions. I happily obliged Cotton by presenting my hand to her for sniffing and then rubbing her behind her soft ears. We sat in Sharon's lovely dining room.

Sharon offered me tea or water and snacks as I set up my laptop, and I declined. Getting herself some water, she said, "I have to have a drink moistening my mouth all the time; my saliva glands were damaged by the radiation." I kept looking at Sharon's face as I arranged my recording equipment and computer, but I didn't want to stare. I had seen Sharon previously at the meetings for the Relay for Life team I had recently joined. I wasn't horrified or disgusted, but I was intrigued. Her mouth seemed to be in the process of caving in—obviously her facial bones and teeth were damaged during the intense radiation and chemotherapy treatments she had more than 30 years before when she was diagnosed with arhinomyo sarcoma in the nasal cavity on the right side of her head. Sharon spoke matter-of-factly, with calm acceptance. At times when she told me painful details, she looked into my eyes as if checking to see if I understood; I felt a warmth in my chest and compassion and empathy flowed from my heart and throughout my torso as I bore witness to a fellow long-term cancer survivor's stories.

Sharon's voice became more urgent, stressing words when she gestured to the photo of the foods she eats – all soft, all with sauces, with lots of beverages to wash them down—on my computer screen, her descriptions of the photos part of the "photovoice" method I am using in this study. Then she continued in a matter-of-fact tone that I recognized from others who have told and retold their stories many times. "During my treatment and then afterwards when the effects of the radiation on my face and neck were so bad, I went from 125 pounds down to 70, because I had mouth sores and I couldn't swallow." I tried to imagine weighing only 70 pounds, but I could not.

I turned to the next photo on my computer screen and gestured to it with a hand wave. "There's a picture of Alfredo sauce and a tray."

Sharon nodded. "That's what I have to eat. Because I'd have a stricture here," she said, gesturing to her throat, "I have to eat stuff with sauce. So the photo shows that I have to eat moist, saucy stuff that goes down easily. I eat a lot of fish. I have to crush my pills, since I can't swallow them. So I take pills that help me develop saliva, and I take them with yogurt." Sharon motions toward the photo again, pointing to the small yogurt carton.

"I just had an endoscopy done. I have to have them every four months. They dilate my esophagus from 8 millimeters up to 14, and that helps me swallow."

"That keeps it open for a while, and then it gradually closes, and then you have to do it again?" I asked. I imagined Sharon having her esophagus dilated over and over again. I started to gag, remembering the endoscopy procedure I had had during my own cancer treatment more than 20 years before.

"Yeah. I'm getting close to it now. The food just kind of gets stuck, at a point where I have to try to jump up and down," she explained. Sharon sat up straight and then slumped her back and shoulders to simulate the process. "Or I lie down to get the food down."

I looked at the photo again, noting how bland all the soft foods were. "Do you still have no sense of taste?" I asked.

"Um hm," said Sharon, nodding. "This is how I describe it. You know how your taste buds are about bitter, sweet, sour, salty? I don't get the flavor of the food, but I get the salt and sugar and bitter and hot. I get the sensations of what's on my tongue, but not taste. I don't have smell either."

"I was going to tell you that the jasmine you have growing by your tree in the yard smells amazing but—", I realized that she could not smell it and felt awkward.

"Oh really? That's why I have candles burning," she said. Turning back to the laptop screen, she presses the arrow key several times, emitting a series of soft clicks, and scrolls down the screen until she finds what she wants. "That's my candle picture. Because I don't have the ability to smell, I don't know what my house smells like. So I put scented candles out to make sure it smells nice."

"Oh, well, it does smell *very* nice," I assured her.

Sharon laughed nervously. "Because I'm not sure, that's why I showed you the candles." I nodded. We continued discussing each photo and her life as a cancer survivor until we came to a wedding photo of a much-younger Sharon in a dress I recognized as the height of 1980s style, and I smiled widely, thinking of my own wedding several years later.

Sharon smiled. "I got married in '84, and then I started teaching in February of '85."

I wanted to hear more about her husband. "Your husband must be a very special person," I offered.

"Uh huh. Yes, he is."

"How did you meet him?"

"I met him," Sharon began and coughed softly, pausing to take a sip of water. "Sorry—dry mouth."

"Oh, that's OK," I responded quickly.

Sharon sipped her water and continued. "I met him when I was out with my sister." We discussed her husband and what they liked to do together, as evidenced in a series of her photos of their motorcycle and small vacation cabin.

Sharon continued, still calm but with a sad expression in her eyes. "Lately, I feel like 'Jim, if you're ready to divorce me, go ahead,' you know. I don't even want to have sex. I feel like I'm a burden on him all the time, because he's always trying to help me eat and care for me and has to deal with me going to restaurants and trying to find foods I can eat. Or you know, how I look. I'm embarrassed of how I look. I've just gone downhill since we got married, and I feel like such a burden to him. I'm sick all the time, and he's always having to deal with me, and he's such a great caregiver. And I couldn't give him a child, although he's said, 'I don't want children. I didn't marry you for children.' He knew. But I feel, sometimes, *worthless*. You know, like 'God, you could have a beautiful healthy wife, and you got me, so you're stuck.'"

My heart aches as I nod sympathetically. I tried to think of how to be kind while not denying or diminishing the pain of what she felt. Hesitantly, I said, "I bet he thinks you have a lot to offer though."

"Yeah," she said, looking down.

I thought of how I felt when my husband took care of me following each surgery—grateful yet ashamed of being so dependent—or how guilty I felt when I kept him up at night when phantom limb pain left me tossing and turning. Looking her in the eyes, I spoke from my heart. "It's not easy feeling like you're the one in the partnership that's a drain on the system. I feel like that too sometimes. Often."

Sharon met my gaze, and we nodded together. We differed in many ways, but we shared some of the same vulnerabilities as women who survived cancer, only to continue to fight the long-term health problems left in the wake of toxic cancer treatments. After sitting in comfortable silence for a few moments, I scrolled to the next photo, which featured her collection of beautiful angel figurines.

Active Interviewing

How do bodies interview bodies? Interviews are a great way to gather information and stories with others. In face-to-face interviews, bodies encounter each other as warm, tangible, messy, material manifestations of ourselves; bodies do not wait quietly outside the room while our "real" essential Self interviews the disembodied mind of the Other. While that may sound obvious, researchers tend to conceive of our interviews primarily in terms of the *words* spoken by our participants, rather than deeply considering the interaction as embodied, emplaced, and constructed. As evident in the narrative above depicting one of my research interviews in a study of everyday communication of long-term cancer survivors, my embodied responses to my participant's photos and her thoughts and feelings reflected our connection as co-members of the long-term survivorship community. I used photovoice techniques—inviting participants to take photos of their lives and then interviewing them individually or in groups to discuss the meanings of the photos (Wang, 1999)—to draw attention to participants' embodied and emplaced experiences through visual representation of bodies (their own and others).

Qualitative social scientists (particularly feminists and other critical, social justice-oriented scholars) have long rejected the mechanistic view of interviews as an extraction of information by a neutral, expert researcher from a vessel-like subject (e.g., Reinharz, 1992). The researcher is positioned not as "a cryptographer or an archeologist . . . [but rather] a facilitator, collaborator, and 'travel companion' in the exploration of experience" (Gemignani, 2014, p. 127). Qualitative methodologists consider interviews to be *active*; that is, two (or more) participants generate meanings within the encounter, through verbal and nonverbal communication, framing interviewing as a co-construction within a specific context. The active interviewer "virtually activat[es] narrative production" (Holstein & Gubrium, 1997, p. 123), eliciting the stories of an interviewee with questions, probes, and comments, but also through physical appearance, facial expressions, gestures, posture, proximity, (possibly) touching, and through voice signals including tone, inflection, speech rate, and volume. The active parties in the interview—both with agency—interact in a site of embodied knowledge construction.

As travel companions or co-constructors, researchers present a persona or role for themselves in interviews, for which interviewees hold expectations; neither party goes into an interview neutrally or without conscious and unconscious goals, expectations, agendas, and emotions. These expectations are socially constructed and culturally situated. Researcher and participants, as van Enk (2009) notes:

> brought expectations to the interviews based on past encounters in other places—expectations about relevant content, acceptable form, and appropriate uptake of roles. Such expectations, as well as the presumed iterability

of accounts, mean that the interview as a context for narrative needs to be conceived of not just as emerging in and out of the "here and now" but also as permeable and historically responsive.

(p. 1271)

Qualitative researchers make conscious choices and also give unplanned reactions that may or may not fit with the conscious effort to project, for example, warmth or a lack of judgment, perhaps a sense of belonging or connection, or conversely a sense of outsider eager to learn from an insider/expert participant. These choices and reactions are grounded not just in the present interaction of body-selves but in past experiences as embodied people. We *do* our bodies in specific ways in interviews, and for example, "whether I 'do' a distantly neutral interviewer or a chummy, self-disclosing one (and in commenting ironically that I'm supposed to be the former, I'm clearly performing the latter), I cannot avoid influencing the words of the interviewee" (van Enk, 2009, p. 1266). Embodied performances are integral to tone and relational communication within an interview.

A researcher who demonstrated awareness of the encounter in which meaning is actively generated is Best (2003), who describes her interviews as a white woman about ten years older than her participants in a study of racially diverse, urban high school prom [formal dance] experiences. As she interviewed two young women of color, she found that they translated their talk as they went along and punctuated their speech with phrases such as "You know what I mean?"

Katy and Serena's . . . acts of translation—their speech acts . . . deployed language in ways that constructed (and sustained) my identity as an outsider and in so doing, discursively sustained my status as "White." Katy and Serena, as they struggled to talk across race, were active in the process of racial inscription. Our identities served as resources to organize and structure this discursive engagement.

(Best, 2003, pp. 903–904)

Whiteness, like other identities, is an "interactional accomplishment;" that is, researchers always *do* race when we do interviewing (Best, 2003).

Mutual performances of self between interviewer and interviewee create intersubjective meanings (Colombetti & Thompson, 2008). Researchers encounter interviewees not just with our minds but with our body-selves. Western cultural discourses, underlaid with principles of positivist science, falsely portray bodies as separate from each other and from the world, as self-contained. More accurately, all sensory experience is interconnected within our mutually constitutive, permeable bodies, such that any sensual practice . . . involves recognizing the other as physically separate yet simultaneously and inescapably connected to us. We understand this as bodies and other elementals of the social being corporally 'inhaled' into us by ontological necessity. Indeed, the concept of *inhalation* captures this experience as grounded in the body. Just as we need to breathe as a biological necessity, we must inhale in order to sustain conditions of subjectivity.

(Riach & Warren, 2015, pp. 794–795; original emphasis)

This means that in an in-depth interaction between interviewer and interviewee, an intensely sensuous interaction takes place. The metaphor of inhaling (breathing) each other helps to convey how enmeshed we become through communication. Rapport is a sensuous experience of bodies sending signals back and forth continually across a connection. When an interview is flowing, researchers may feel connected to our participants and forge and reinforce that connection by leaning forward (thereby becoming closer to the participant), making extended eye contact, perhaps by touching an interviewee's hand or forearm. We also feel the absence of connection with our bodies; our shoulders stiffen when an interviewee crosses their arms over their abdomens in defiance or defense, our stomachs sink when our interviewee gives only monosyllabic responses, our faces fall in dismay when we realize we have inadvertently offended our interviewee or probed too hard about a painful topic. We can tell some of how the other person is thinking and responding because we have been "inhaling" each other and getting a feel for how we fit together (and perhaps some sharp edges where we do not fit).

Moreover, as with any interaction, the meaning of interviews remains in flux rather than fixed (Blumer, 1954), and interaction may vary significantly over the course of an interview. Interview conversations vary in flow, tone, and quality (Corbin & Morse, 2003). Yet researchers tend to describe interviews as good or bad; we "clicked," or we failed to connect. This may be misleading. Most interviews start off slowly, all have peaks and valleys, and sometimes the most enlightening statements and useful responses are made following a "clearinghouse" question at the very end of the interview (e.g. What else is important to know about this topic? What haven't we talked about yet on this topic that you think I need to know?) and the more informal wrap-up talk. In emotionally fraught interviews, crying or feeling sadness tightens the throat or facial muscles as one's heart aches with the participant's pain, and feeling physically and emotionally drained afterward is to be expected. At the same time, interviews can also leave researchers feeling happy and energized, with great insights and the pleasure of enjoyable interaction.

Language of the Body: Understanding Nonverbal Communication

Qualitative studies tend to emphasize what participants said (i.e., their language) as the basis of meaning and incorporate little or no explicit attention nonverbal communication (i.e. body language) within their conventionally reported findings (i.e. journal articles). Denham and Onwuegbuzie (2013) reported that qualitative researchers either do not collect nonverbal communication data, or they collect it but do not analyze or interpret it. On reflection, I realized that while I do collect nonverbal communication data—in the form of interview notes or fieldnotes of observations—I generally draw on such data implicitly to help me interpret participants' words but do not explicitly incorporate nonverbal communication data into my findings (with the exception of (auto)ethnographic narratives, discussed in Chapter 8). Yet nonverbal signals are crucial to the meanings generated within interviews. Signals such as tone of voice and eye contact are rooted in the body; interviewer and interviewee adjust their communication from moment to moment in response to each other's cues. Researchers ignore this information at our peril, as communication research shows that the significant majority—estimates vary between 93% on the high end and 60–70% on the lower end—of meaning manifests not in *what* is said, but in *how* it is said, i.e., nonverbal communication or embodied signals that are commonly referred to in popular discourse as body language (Hickson, Stacks, & Moore, 2004). Participants are well aware that meaning is constituted in nonverbal signals; Manning's (2015b) study of coming-out narratives of lesbian, gay, and bisexual people reports positive and negative communicative behaviors, including nonverbal signals such as laughter, joking, and nervous habits. Of course, some researchers do attend to nonverbal communication; in an analysis of qualitative articles published between 1990 and 2012, researchers identified the following purposes for including nonverbal communication in reporting of qualitative findings: clarification, juxtaposition, discovery, confirmation, emphasis, illustration, elaboration, complementarity, effect, and corroboration/verification (Denham & Onwuegbuzie, 2013).

Becoming more aware of nonverbal communication signals—your own and your participants'—can help enrich understanding of participants and their meanings. People are always communicating (at least subtly) with those around us, and even when alone. Most of the ongoing communication of who we are, how we feel, and what we are doing is nonverbal, often without our conscious thought, although we can make strategic choices in our nonverbal communication as well. Especially for those outside of the communication discipline, a review of some basic nonverbal communication categories and terms offers a richer vocabulary through which to attend to, document, and interpret embodied cues.

Nonverbal communication can be defined as the process of intentionally or unintentionally "creating meaning in the minds of receivers . . . by use of actions other than, or in combination with, words or language. Nonverbal communication includes norms and expectations . . . for the expression of experiences, feelings, and attitudes" (Hickson, Stacks, & Moore, 2004, pp. 14–15). Nonverbal communication performs six functions: identification and self-presentation, control of the interaction (i.e., regulate flow of conversation), indicating the type of relationship between communicators, display of cognitive information (e.g., OK, Stop!), display of emotions, and display of deception (Hickson et al., 2004, pp. 21–22).

Qualitative methods textbooks offer little or no coverage of collecting or analyzing nonverbal communication (Onwuegbuzie & Byers, 2014), although texts on discourse analysis explain systems of codes for indicating vocal cues such as emphasis, tone, hesitations, and nonlexical sounds in transcripts (e.g., ugh, hmmm) (Edwards, 2003). Breaking nonverbal communication down into types may be helpful in coming to understand the degree to which meaning is a full body expression rather than primarily about words. Keep in mind that while these signals can virtually all be consciously enacted, much of their display seems automatic and goes unnoticed because we have been socialized into norms, and our own personalities are grounded in well established patterns of expressing ourselves with our bodies. I offer a brief explanation of different types of nonverbal communication so that researchers can consider separately and analytically what we usually process subconsciously, intuitively, and extremely quickly. By looking at each type separately, I hope to expand possibilities for noticing cues and bringing them to conscious awareness during interviews (and participant observation).

Kinesics includes body orientation, posture, and gestures; this category includes most of what is commonly referred to as "body language" (Hickson et al., 2004). Relevant issues for interviewers include how one's posture is when meeting an interviewee—uncrossing the arms indicates openness, erect but relaxed back and shoulders reflects a degree of comfort, a smile conveys welcome. Face and eyes signals are very complex and can be quite difficult to interpret (Ekman, Friesen, & Ellsworth, 2013). Further, interviewees might wave their hands for emphasis while telling a story, play nervously with a pen or the shoulder strap of a purse, smile when detailing their children's names and ages, and sniffle or cry while describing the death of a loved one. Keep in mind that kinesthetic signals can be as small as a barely discernable widening of the eyes or as large as waving both arms vigorously.

Proxemics refers to how much space people leave between one another (Hickson et al., 2004). Many social rules dictate appropriate distances, and norms vary widely by culture. Generally, in Western cultures, intimate distance is less than eighteen inches, and personal distance is considered to fall between eighteen inches and four feet. Social distance ranges from four to twelve feet, and public distance is greater than twelve feet and usually involves giving a presentation to an audience. When there is no space between people who are touching, that is referred to as "zero space" (Hickson et al., 2004, p. 19). In an interview, how close the interviewer positions the chairs, and whether there is sufficient space for the interviewee to choose to draw closer or move further away, will influence the communication between them. Proxemics are constantly changing as the interview progresses, at least in minor ways, as interviewer and interviewee lean forward or back as they talk. Interviewer or interviewee may lean forward excitedly to convey passion for a topic or lean toward the other with empathy for another's suffering, decreasing the social space between them as they seek connection, increasing space in discomfort, or just seeking to adjust the social space to their own comfort levels. In a walking interview (discussed later in this chapter), space will fluctuate even more substantially, and negotiations of space always reflects and influences ongoing relational signals between interviewee and interviewer. Proxemics is one of the nonverbal communication signals that tends to be noticed only when implicit rules for it are violated, but it is always a crucial part of interaction, and sitting too close, too far away, or at a mutually comfortable distance all communicate to the interviewee about the interviewer and the purpose and expectations for the interview.

Vocalics, or paralanguage, focuses on how we use our voices, beyond just the articulation of specific words, to convey meaning (Hickson et al., 2004). People are socialized and learn unconsciously that others respond to how loud or soft we speak, how fast or slow, and how we speak with intensity for emphasizing a specific word or phrase. Tone of voice may convey a range of meanings that tell the listener how to interpret the words being said, such as sarcasm, compassion, or joking. Disfluencies include nonlexical sounds (not words) such as uh, ah, um and repeating or stumbling over words, which may indicate nervousness, uncertainty, or such intensity of emotion that the person is overcome by it and struggling to speak. The sound "mmm" may convey pleasure, for example while eating (Wiggins, 2002). Young children are schooled in how to convey politeness of tone (respectful, submissive to authority), as well as words (please and thank you). Interviewers generally try to infuse their voices with a welcoming tone of being pleased or happy to conduct the interview, for example. Interviewers may use a tone infused with kindness and gentleness when discussing a painful topic. Crying, snorting, grunting, sighing, and laughing are forms of paralanguage as well. Laughter sparked by the interviewer or the interviewee may evoke a response of shared laughter, or when the laughter is rueful, expressions of sympathy or kindness, or if the laughter is self-deprecating, expressions of ironic agreement or polite disagreement (van Enk, 2009).

Haptics means touching (Hickson et al., 2004). Because humans are relational, human touch is essential to healthy development in human babies (Field, 2014). Touch expresses relationship status of those interacting. A handshake is a social ritual that may signal the beginning or end of an interaction between friends or acquaintances, for example. Appropriate touch with strangers or acquaintances varies significantly by culture. Thus an interview could commence with a greeting between interviewer and interviewee that ranges from embracing to kissing on both cheeks to a hand shake to bowing toward each other without touching to a verbal greeting only. Hugs are exchanged between close friends and loved ones but also casually between friendly acquaintances or even strangers in some contexts. A light touch on an interviewee's arm may reflect sympathy, an attempt to get someone's attention, or warmth and liking. Of course, touch occurs in conjunction with facial expressions, changes in proximity, and other gestures that come together to convey meaning. Physical violence is a form of touching that can range from a soft slap to dangerous hitting or beating. Touch also conveys sexual interest (and lack thereof).

Chronemics involves the use of time (Hickson et al., 2004), including calendar time, personal experiences of time (e.g., as fast or slow), and biological time (individual biological clocks), and informal and formal time, among others. While most of us think of time as an inevitable, measurable reality, our understandings and experiences of time are socially constructed through communication with others (Gergen, 1994); time "is fundamentally an *intersubjective* phenomenon" (Ballard & Seibold, 2004, p. 2; original emphasis). Perceptions and experiences of time act "as important symbolic nexes around which coalesce issues of order, power, self-definition, and knowledge" (Dubinskas, 1988, p. 3). Setting aside a designated time for an interview, whether pre-arranged or spontaneous, communicates on both sides some degree of willingness to interact, ranging from reluctant agreement to enthusiasm. Being late for an appointment may communicate disinterest or a lack of respect. How long one allows an interaction to go on also indicates interest and also a degree of respect or possibly liking.

Physical appearance is often considered to be embellishment of the body through clothes and accessories (hat, purse, backpack) and grooming (e.g., hair style, make up, tattoos, body cleanliness), and (to some degree) body shape (height, weight, muscle tone) (Wood, 2012). Theorist Goffman (1959) referred to all such embellishments as an individual's "personal front." From a communication perspective, the choices we make about fashioning our appearances, as well as features that we generally do not choose, such as race, all constitute nonverbal communication to others about our identities. Clothing conveys messages, and how dressy or casual, (in)expensive, (un)revealing, (non) trendy, (non)conforming with social norms our clothes are tells people who we are, although not always as intended. Perceived race, gender, class, sexuality, approximate age, attractiveness, and (dis)ability—as (mis)judged by our appearance—also quickly communicate messages to others about the individuals with whom they interact. As people get to know each other, physical appearance matters less, but it remains fundamental to how we understand others. In interview studies, such "issues related to self-presentation and what aspects of a researcher's identities are highlighted (or backgrounded), play an important role in . . . the process of knowledge production" (Razon & Ross, 2012, p. 495). For example, Del Busso (2007) considered some of the ways that researchers' bodies communicate, sometimes in ways we do not intend, when bodily signifiers, such as grooming and clothing, are interpreted by our participants. She explained, "I felt that my body had spoken directly or indirectly to the women who took part in the research in ways that were incompatible with my feminist identity" (p. 311); that is, her participant did not think she "looked" like a feminist because she was "too feminine" in appearance. Often such perceptions relate to stereotypes and the degree to which researchers match or do not match their interviewees in demographic categories (e.g., age, race, sex) but also in terms of beliefs, values, and life experiences. Wearing a cap or t-shirt with a local sport team's logo on it, for example, can bring people together or spark a disagreement based on competing loyalties, becoming part of the interaction in either case. At times marks on the body communicate to others: Harris' (2015) study of heroin users involved her embodied identity and physical appearance as a recovering heroin addict. During some interviews, "my body disclosed my past—a decade of injecting excesses writ large on the arms. These disclosures, embodied and verbal, affected the energy of the interview dynamic and, in turn, my embodied understandings of illness and drug use" (Harris, 2015, p. 1689).

Territoriality refers to the configuration and decor of the physical environment or place associated with a person or organization, such as an office or bedroom (Hickson et al., 2004). The interview location and social context are integral to communication within an interview (Gemignani, 2014). Context is not just "out there" as a backdrop; it signals meaning about the researcher and the purpose of the interview. The objects that occupy our territories and their arrangement—in our homes, offices, even lockers at school or at gym—communicate about our identities and personalities, not always in ways that we might desire or control. For example, I have conducted interviews in my university office. This crowded room contains some things that might be expected of a professor's office—many books, a computer and desk, harsh fluorescent lighting, photos of my spouse and cats—but it also contains a small biplane constructed of empty Diet Coke cans, a poster of the "Periodic Table of the Desserts," a calendar showing paintings of scenes from the coast of Maine, a tall pen and holder shaped like a flamingo, a comfortable stuffed chair for visitors, and a small refrigerator covered in pictures of my nieces and nephews surrounding a bumper sticker that says, "Feminism is the radical notion that women are people." When I interview people in this space, the atmosphere or feel of my office is integral to our communication. If I met interviewees in a sterile office devoid of personalizing features, that too would impact their perception of me, the institution that employs me, and possibly the disciplines I represent (communication and gender studies). When we interview people in the contexts in which they work (e.g., clinic, office building, restaurant), in their own home, or outside in a park, all of these contexts unavoidably generate meaning as part of the interview.

In summary, the many types of nonverbal communication offer a vocabulary for identifying types of signals through which people (intentionally and unintentionally) convey meaning. Although I have addressed types separately here, in practice, many nonverbal cues are used simultaneously, in endless combinations, during interactions.

Power and Resistance in Interviews

Power and resistance are experienced in and through our bodies and those of our participants. Corbin and Morse (2003) explained that the issue of power in research interviews is complex; it is not a simple matter of participants being exploited by researchers for the researchers' own professional gain, nor does the interviewee gain nothing from the interview process. Instead

> [t]here is reciprocity between researcher and participant . . . with each gaining something from the experience. In addition, there are benefits to society in the form of professional knowledge development. These benefits are not undervalued or overlooked by participants, who often consider the opportunity to participate in research as the opportunity to give back indirectly to society.
>
> *(Corbin & Morse, 2003, p. 349)*

Moreover, participants receive the experience of researchers' embodied participation in the interview and witness to their stories. Both participants exert power within the interaction, and although the interviewer arguably has more power, interviewees are not without the ability to assert control and to enforce an agenda of their own. Power and resistance can be experienced in an interview interaction in a myriad of potential material ways. A commanding or formal tone of voice or signals intended to portray the image of an authoritative expert, along with wearing a suit or formal dress, may project a professional, interested-but-detached persona, while a warmer, friendlier, and less formal greeting and attire would offer more subtle signals of power.

Choices can be made by interviewers to share power to greater or lesser degrees. Mainstream approaches to interviewing define the interviewer as being in a position of power and control, while the interviewee has little or no power (Mies, 1983). Some methodologists recommend seeking to minimize that power difference (Lindlof & Taylor, 2010). Ezzy (2010) suggests that interviews can be framed as more like communion or more like conquest. Qualitative methods textbooks descriptions of interviews are "replete with masculine metaphors of conquest: probing, directing, questioning, active listening. . . . the interviewer . . . is in control, directing and shaping the course of the conversation" (Ezzy, 2010, p. 164). She advocates a communion model that involves more egalitarian discussion based on mutual interests, offering and sharing rather than questioning or demanding. Ideally, interviewer and interviewee become mutually interdependent, giving to one another through the exchange so that the conversation becomes a gift to both. In this way, researchers and participants foster connection, "a moment of recognition of simultaneous sameness and difference" (Ezzy, 2010, p. 164). There can be room for probing as well, as long as it is done with care. For example, "active probing" versus "attentive openness" reflect different emotional framings of interviews. These models are more of a continuum than a dichotomy, and the ways in which they differ are grounded "as much in the embodied, performed, and emoted experience of being together as it is in the detail of the meanings conveyed" (Ezzy, 2010, p. 165). That is, nonverbal elements of communication contribute significantly to how power is performed, experienced, and resisted.

Feminist and other critical researchers have long advocated sharing power more equitably between researchers and participants in interviews. Because research participants tell their own stories in their own voices (rather than responding to survey scales with numbers or participate in an experiment controlled by a researcher), the interview structure can generate the research participant's power as well (DeVault, 1990; Mies, 1983). Broad, open-ended interview questions allow research participants to control the direction of the conversation to a greater extent than do closed-ended questions (Mishler, 1986). Participants also make choices. They can choose what to reveal and conceal, when to be truthful or not. "[R]esearch participants use silences and fragmented stories as strategies for countering power differentials and oppression" by researchers; they may actively engage in resistance (Razon & Ross, 2012, p. 502). Ideally, in qualitative research an interactive conversation is created between participant and interviewer that provides plenty of room for participants to engage in sense making about issues that are important to them. Some researchers believe focus group interviews enable more equitable power sharing because the participants share communication with each other as well as with the facilitator (interviewer): "One of the chief attractions for me in using focus groups was their non-hierarchical structure, the lateral transfer of power to participants consistent with my feminist ethic and the equality sought in organic inquiry" (Moloney, 2011, p. 68).

Moreover, it is a mistake to think of power as fixed within an interview. Power is always shifting, multiple, and contextual.

> [C]learly power also intersects with a range of emotional registers displayed throughout an interview encounter. It is through increased attention to these *shifting* dynamics of power, and the *intersections* of power, that redefinitions of the research relationship can take place—illuminating the existence of fluid identities, multiple subjectivities and mobile positionings.
>
> (Hoel, 2013, p. 45, original emphasis)

For example, researchers interviewed learners about their educational experiences. Yet, they explain that interviews reflected "plenty of flux and uncertainty about what it means to tell one's educational story to a researcher in an interview. And this flux and uncertainty draws in again the processes of negotiation that take place in conversation" (van Enk, 2009, p. 1273). Much of that flux and uncertainty is communicated through embodied cues, that is, nonverbal communication.

Re/Membering the Body

Even under ideal circumstances, participants may find it difficult to express the mundane, largely taken-for-granted aspects of everyday, sensory experiences.

> [T]he elusiveness of the lived ordinary . . . [is] particularly marked in the case of the research interview. If interviews are . . . a means of making sense by inviting respondents to do the same, then the very first lesson of reflexivity is that we should acknowledge the full implication of this asser-tion. Interviews *make* sense; they fabricate meaning.
>
> *(Hall et al., 2008, p. 1023; original emphasis)*

That is, interviews become the impetus for participants to think explicitly about abstract, largely ignored concepts and embodied experiences. Thus researchers inquiring about embodiment in daily life are not accessing participants' *prior* meanings but participating in participants' *creation* of meanings about embodied perceptions and practices. These constructed meanings can offer great insights into participants' embodied worlds.

Moreover, interviewers can ask not just about sensory experiences but use "the senses as a portal to experience more broadly, particularly to experience what is otherwise too difficult to articulate, too mundane to recount, or too intangible to otherwise access" (Harris & Guillemin, 2012, p. 692). For example, each of the senses can make us remember different times, places, people and experiences, and researchers can prompt interviewees to recall sensory details.

> [I]n our everyday lives, particular smells can force us to instantly recall people or moments in our lives; touch can evoke strong memories and feelings; sounds, such as music, can take us back decades in time; and taste induces a spectrum of embodied reactions from sheer pleasure to strong disgust.
>
> *(Harris & Guillemin, 2012, p. 690)*

Sensory memories are, of course, situated in the specific times and places in which they are co-constructed in interviews. Further, participants' and interviewers' co-construction of past or remembered bodies is influenced by discursive lenses through which we (re)envision the past.

> [T]he past [is] a political achievement as well as the result of individual practices of remembering . . . [T]he creative act of remembering situ-ates the teller in relation to his or her memories and also in relation to the listener and larger contexts of discourses, collectivities, and possible constructions.
>
> *(Gemignani, 2014, p. 131)*

The remembered body may be accessed through interviews in a variety of ways. Råsmark, Richt, and Rudebeck (2014), in a study of habilitation therapists, asked therapists to discuss using their bodies to work with children with significant physical disabilities. The therapists readily reported some ways they used their bodies, commenting on these embodied memories in a "vivid and committed way," (¶56), which the authors interpreted as evidence of their passionate, moral commitment to caring for the children. Habilitation therapists also expressed surprise and new awareness of other aspects of their embodied practices. By inquiring about the therapists' bodies, "the interview situation as such, by a mere refocusing of memory, furthered body awareness, in turn making a level of latent body experience available" (Råsmark et al., 2014, ¶58). Such increased body awareness, the authors suggest, generates attention to professional practice and increased capacity for empathy with the children's suffering and triumphs.

To Match or Not To Match—Performance of Shared Identities

Professional perspectives on the gender of the ideal interviewer have evolved over time (Warren & Karner, 2009). Famed sexuality researcher Alfred Kinsey and others steeped in the positivist model of science (as well as structural sexism and racism) believed that white men could be trained to be the best interviewers because of their "natural" neutrality. With the cultural shifts in the 1970s, (white) women became seen as "naturally" nurturing, empathetic relational specialists who could elicit more information out of participants. In hindsight, these type of assumptions based merely on biological sex and race seem unwarranted and even silly, yet assumptions about natural abilities remain pervasive and fuel stereotypes. But while the notion of a singular "ideal interviewer" type of body has been largely dismissed, the notion that there ought to be a "matching" of interviewers and interviewees based on demographic categories, key life experiences, or both remains a topic of ongoing dialogue—e.g., that women should interview women, gays and lesbians interview gays and lesbians, indigenous people should interview indigenous people, alcoholics should interview alcoholics, and so on (e.g., McDonald, 2013). There are both strengths and weaknesses to this approach to organizing the bodies of interviewers for specific types of participant bodies.

On the plus side, when researchers and participants collaborate on the basis of common perspectives (presumably grounded in shared racial, gender, or other embodied identities and experiences), the researchers may be able to better help participants to tell a story by responding in ways that indicate understanding and empathy, referred to as "congruity" between interviewer and interviewee (Riessman, 1987; see also Gunaratnam, 2003). Not just categorical similarity but perceptions of particular embodiments—such as women with disabilities or male athletes—may lead participants to feel more comfortable with disclosures. One female researcher with a disability wrote about interviewing other women with disabilities. "Perhaps my own body compelled these particular narrators to disclose these experiences. Our similar embodiments engaged in the performance of teller and listener reciting cultural performativities familiar to each of us" (Scott, 2015, p. 230). There is good reason to believe that shared understandings can lead to better communication within an interaction. However, it is less clear how rooted shared understandings are in shared physical characteristics and the shared socialization and stereotypes that people are subjected to on the basis of those characteristics, as compared to how rooted they are instead in compatible communication styles, and the affording of mutual respect, sincerity, and openness by both interviewer and interviewee.

On the other hand, the main disadvantage to matching is that it is impossible to match based on all or even most categories. Matching various aspects of difference is always intersectional, and no one can control for all differences; "Perfect congruence . . . is probably not possible, not even always desirable" (Riessman, 1987, p. 191). Also, it is difficult to know which differences are the most significant ones in a given situation (Carrington, 2008). Since many categories are not typically disclosed, there may be significant differences of which researchers and participants are unaware prior to the interview (McDonald, 2013). Further, "'[m]atching' for embodied categories of difference is not necessary because some research accounts cannot be seen as inherently better than others simply on the basis of researchers and participants identifying with the same identity category," especially since all identities are fluid (McDonald, 2013, p. 139).

Moreover, the belief that people of one group will not be as open with those of another, such as women when being interviewed by men, or people of color when being interviewed by white people, is often mistaken. Hoel (2013) pointed out that among her participants, given a choice between being interviewed by a Muslim woman of Indian descent (like the participants) or by the researcher who was a non-Muslim, Norwegian, white woman, most chose the researcher. Perhaps this was because "a non-Muslim western researcher might be perceived as less likely to impose value judgment on respondents' experiences and decisions that they made within intimate relationships, particularly if these decisions jarred with dominant understandings of Islam" (Hoel, 2013, p. 32). Likewise, in a study of masculinity in workplaces, an all-female researcher team studied male firefighters, and participants reported that they were more comfortable discussing emotions and concerns with the female researchers than with their male colleagues (Hall, Hockey, & Robinson, 2007). Allen (2005) (a woman much older than her male participants) contended that she listened with sensitivity and "genuine interest" to her participants' discussions about sexuality, creating a comfortable environment "which was more conducive to the emergence of less traditional male sexualities than these qualities being a direct consequence of simply being female" (Allen, 2005, pp. 51–52). Allen's experiences reinforce the importance of communication and interviewing style relative to that of demographic similarities in establishing rapport and a sense of security in disclosing.

Furthermore, embodied identities of researchers and participants are most usefully thought of as fluid within interviews:

> [D]ifferent parts of identity [are] . . . negotiated, highlighted in some points, or discarded all together in particular moments . . . [W]e therefore employ the term *fluid identities* to capture the flexible, overlapping, and at times conflicting identities researchers and participants hold. We use fluid to signal movement rather than bounded-ness of insider/outsider dichotomies or hybrid notions of identity . . . [and] as a way of capturing the way that our research participants categorize us, and our choice, in turn, to accept or challenge those categorizations.
>
> *(Razon & Ross, 2012, p. 495)*

Neither interviewers nor interviewees can be certain that the personas they intended to project were interpreted as intended, or that having been recognized as intended such identity claims were accepted. Fluidity of embodied identities relates to any number of factors, including the embodied experiences of interviewer and interviewee—e.g., pain, hunger, fatigue, excitement—during the course of the interview. In addition, as elements of experience and meaning are revealed, they may change what we think we know about others and our responses to them.

Significantly, while participants' similarities to the researchers' (past) bodies may cause emotional responses that affect participants' disclosure in interviews (positively, negatively, or both), similarities also may resonate deeply within researchers' bodies. A recovering heroin addict studying hepatitis C in addicts, for example, described instances during interviews when "my muscles tense in reaction to the sight of a participant's bruised and scarred arm or my gut will clench upon hearing or revisiting a certain story. . . . awaken[ing] bodily memories of my drug-using years" (Harris, 2015, p. 1696). These similarities may lead to empathy and greater understanding but also to pain and possible risk in this case. As a cancer survivor who studies cancer patients and other patients with serious illnesses, I often have experienced visceral memories in responses to participants' stories.

Technology, Mediated Bodies, and Embodiment

Bodies can be and often are mediated in research processes. In this section, I briefly discuss embodiment in interviews that are conducted via telephone, the Internet, or smart phones and using teleconference software, instant messaging, email, or specialized apps. A full exploration of mediated interviewing via phone (see Shuy, 2003) or Internet (see Fielding, Lee, & Blank, 2008; James & Busher, 2012) is beyond the purview of this book. Yet I want to address briefly that while interviews may appear to be disembodied by technology, researchers' and participants' bodies present and are implicated in complex ways. Researchers engaging in mediated interviewing have to learn how to demonstrate their taken-for-granted ways of establishing rapport face-to-face with participants in online exchanges and how to read participants' nonverbal cues. "Certainly online data collection obscures the more overt attributes of the conventional research site, but the setting, like the body, does not necessarily disappear in cyberspace" (Seymour, 2007, p. 1194). Mediated contexts enable a different form of togetherness—virtual, often asynchronous, technologically mediated—than that afforded in face to face interactions.

> Physical co-presence is no longer the only presence possible. There is presence in absence, privacy in public and connectivity, even intimacy, in isolation. . . . Our senses of place, home, self, identity and of our own bodies are being re-negotiated, shifted and at times suspended.
>
> *(Enriquez, 2012, p. 60)*

And yet face-to-face and mediated communication are not dichotomous or firmly delineated from each other with the former fully embodied and the latter embodied not at all or only by inference. "Rather than existing as separate realms that the body must work between, the virtual and the material are instead collaborative spaces that produce embodiment" (Farman, 2015, p. 105). That is, experiences of communicating online—in virtual presence of others, including interviewers and interviewees—contribute to embodied identities and experiences of embodiment both online and in-person. For example, researchers now use e-mail interviews (Bowden & Galindo-Gonzalez, 2015; Cook, 2012: James, 2007; Murray, 2004). I have used teleconference software to conduct photovoice interviews over the Internet, enabling me to see participants (Ellingson, 2017).

Phone interviews have long been employed as a mediated interview strategy (Burke & Miller, 2001). Phone interviews are often (implicitly or explicitly) considered inferior to in-person qualitative interviews; one may miss participants' visual cues, have more difficulty in forming rapport, and not share the same context with participants, which may also lead to a loss of valuable contextual data (Novick, 2008). Often cited in terms of efficiency, phone interviews also may provide more scheduling flexibility, decreased costs, increased geographic range for recruitment, and increased researcher safety, while enhancing interviews by increasing perceived confidentiality, protecting interviewee's privacy (Drabble, Trocki, Salcedo, Walker, & Korcha, 2016). Telephone interviews may even be considered participant-centered as they can enable more comfortable, and potentially more honest, disclosure of sensitive or stigmatized information without the intensity of face-to-face interaction, especially as people become more accustomed to online communication (Trier-Bieniek, 2012). Johnson and Quinlan (2016) provided an interesting twist on the benefits of phone interviewing in their study of communication in an infertility clinic. During data collection, one of the researchers became pregnant, and as her pregnancy progressed, the researchers decided to have her conduct phone interviews only, because her pregnant body may have been painful to participants and discrediting to her as a woman whose embodied state was the very one sought by participants at the clinic being studied.

Practices for Embodied Interviewing

Be Embodied in (Some of) the Moment(s)

To the extent possible during and immediately following an interview, attend to brief moments in which your awareness of connection or disconnection is/was heightened during the interaction, reflecting on how your body-self and your participant's body-self were intersubjectively producing each other and the interaction. "How your own sensuous body responds at a particular temporal moment, experiencing convergence, conflict, or despondency; these are times at which the body speaks . . . [;] where precarious bodies interact in unfolding processes that are ultimately about embodied being" (Hoel, 2013, p. 41). Develop the habit of making notes not only on what you see and hear but also on sensations felt throughout your body in moments of connection or mutual understanding with participants—muscle tension, bursts of joyful energy, heart-aching sadness, and so on, for later reflection and consideration. Learn to recognize such moments and to attend with deliberate intent to embodied sensations, emotions, and insights in the moment that they happen, reflecting on them later.

Keep Quiet and Actively Listen

Listening attentively is difficult, bodily work; even experienced researchers may be challenged to stay out of the participants' way and let them talk while offering appropriate nonverbal cues of attentiveness (Ezzy, 2010). Bearing witness to others' experiences, particularly of suffering, as a respectful and caring witness is a profoundly ethical act that helps people to construct livable narratives for themselves (Frank, 1995). As feminist theorist Luce Irigaray explained, "To be silent to allow you to speak, to give birth to you. And to us as well" (Irigaray, 2001, p. 14). Empathy through nonverbal communication is vital yet mostly silent. There may be crucial differences between researchers' and participants' experiences, yet there is also commonality and connection based in human suffering, such that, when sharing

> intense and distressful moments, researchers often connect with participants at a very deep level. . . . and provide the empathy and support that participants might need to work through troubling experiences. Much of this is done in silence, the researcher sitting, being there for the participant.
>
> *(Corbin & Morse, 2003, p. 343)*

While all research interviews do not have to have therapeutic goals, the gift of listening should be given to all participants, regardless of the research topic. Eager for clarifications, to cover a range of topics and elicit rich descriptions, researchers may jump in too quickly to guide the conversation. Become more comfortable with silences and concentrate on bearing witness as participants fill the silences in the process of making sense of their experiences out loud, in the moment of telling.

Develop Questions That Elicit Embodiment

Develop your interview schedule (i.e., your list of questions) in advance for formal interviews. Plan questions that ask about not only about what participants see and hear, but also about touch, taste, smell, texture, temperature, and movement in the contexts being explored, such as workplaces (Riach & Warren, 2015), homes (Pink, 2005; Pink, Mackley, & Moroşanu, 2015), and neighborhoods (Hall et al., 2008). Ask about different areas of the body—limbs, head, belly, lungs—and perceptions of movement—walking, standing, lifting/carrying, sitting, and other relevant bodily practices. In both formal and informal (i.e., ethnographic or spontaneous) interviews, respond to participants' stories by asking for memories of how their bodies felt in those moments and in what part of their bodies they felt their emotions.

Body-Anchored Interviewing

Consider trying (or adapting) the strategy of "body-anchored" interviewing. Stelter (2010) devised a method for helping participants to describe embodied experiences that are not typically expressed through language—possibly about athletics, dance, or work practices. First the researcher invites the participant to relax and to try to put herself or himself into a moment of the "felt sense" of the issue to be explored. Then the participant is directed to stay with that moment until a word or image emerges, which Stelter terms a "handle." Following that, the researcher and participant work together to consider whether the word resonates with the felt sense and discuss the felt sense and possible words together until the participant is confident that they have arrived at suitable word or phrase (for more details see, Stelter, 2010). This method of elicitation provides more guidance for asking about embodied experiences that is useful for people who are not accustomed to explaining in language what they accomplish through their bodies (e.g., how they do a penalty kick in soccer or change a tire on a car).

Collect Nonverbal Communication Data

Most obviously, when conducting interviews, researchers should take detailed notes on their own bodily reactions before (e.g., anticipation, nervousness), during

TABLE 5.1 Sample Nonverbal Communication Data Collection Form

Nonverbal Communication	
Kinesics	Chronemics
Proxemics	Appearance
Vocalics	Territoriality
Haptics	Other

(e.g., tension, pleasure, curiosity, fatigue), and after (e.g., energized, drained, excited, confused) interviews. In addition, make detailed notes on how participants talked, what they wore, and characteristic gestures, including any nervous habits or twitches. Researchers might also collect nonverbal communication data via the use of audio or video recording. Other forms of data collection include using checklists, matrices, or Venn diagrams that enumerate nonverbal signals (Onwuegbuzie & Byers, 2014, p. 195). For those interested in a set of highly specific guidelines, you can learn how to integrate nonverbal data in a study using a thirteen-step process for collecting, analyzing, interpreting, and reporting nonverbal data in three iterative, recursive phases (Onwuegbuzie & Byers, 2014). For a more informal guide, use the list of categories in Table 5.1 to organize your notes as you observe participants.

Attend to the Absent and Untold

Whatever is made present as part of the interview process necessarily implies the absence of other information and perspectives. Thus what we attend to as material is directly tied to that which is *not* encountered as material.

> Presence *implies* absence. . . . [I]t's a problem when we *imagine* or *pretend* that everything can be made present and known by the all-knowing subject, the all-seeing eye, or the all-representing database. This can only be a pretense because . . . the knowable is dependent on, related to, and produced with the unknowable: that which is elsewhere and absent.
>
> (*Law, 2007, p. 600, original emphasis*)

Thus in our efforts to detail the clothes worn and to attend to the participant's body, for instance, we must remember that each of the details we record implies details not told; each identity discernable suggests one that cannot be displayed, and everything said points to the absence of that which goes unsaid, that is, "all narratives tell one story in place of another story" (Cixous & Calle-Gruber, 1997, p. 178). On a practical level, that means reflecting after interviews about what (you think) you know about participants' body-selves and their embodied experiences and reflecting on questions unanswered and topics not explored. Remain aware that all knowledge of the body and sensory experiences is partial.

Walking Interviews

Walking interviews are another possibility for exploring embodiment which

> draws participants into the tactical, the movement of bodies through space, in time, the negotiation of paths and unforeseen interruptions—and this material wandering encourages the metaphorical wandering of thought, the expression of affect, such that what may not find proper expression in the visible-strategic, finds voice.
>
> *(Kuntz & Presnall, 2012, p. 733; see also, Riley, 2010)*

Walking interviews spark conversations that differ from stationary ones; while not better or worse, they are likely to vary significantly, even when the same questions are asked or topics addressed. Sense making and speech happen not because of, or in spite of, researchers and participants walking, but *through* their motion of walking together as body-selves. Interviews could also use the ever-changing context of the places, people, and things that they pass while trying to consciously become aware of and inquire into ambient noise, vibrations, and sensations, a technique referred to as "soundwalking" (Hall et al., 2008). Generally, the relationships among the contextual elements and the participants' embodied experiences may be articulated when attention is focused on how the background functions in their lives.

Another potential benefit of walking interviews, or "touring interviews," is how the ambient noise of public spaces may act as a power equalizing device, illustrating again the relevance of context for interviews.

> Out and about, the pressure is off. It is much harder—intentionally or otherwise—to put one's interlocutors on the spot, to oblige them to deliver (themselves). Questions asked in noisy spaces lose the authority a quiet room affords, thinned out as they are by the myriad other sounds into which they dissolve. Answers, likewise, bear a lesser weight of expectation: respondents can speak without breaking silence; they can pause without things going quiet.
>
> *(Hall et al., 2008, p. 1035)*

While walking interviews may not always be practical or desirable, when this strategy is possible it offers a way to link embodied movement and speech in a creative, generative manner. While Hall and collaborators walked around neighborhoods, other options might be walking interviews in a large organization being studied (e.g., school, hospital, convention center, museum), or even just inviting the participant to walk in a local park or walking path. Of course walking interviews necessitate a clip-on microphone for both interviewee and interviewer.

Object-Interviews

Incorporating material objects into interviews can prompt discussion of topics that may not otherwise have been addressed, or that would have been addressed less richly. Focusing on the presence of actants in participants' lives (Barad, 2007), Nordstrom (2013) developed what she called an "object-interview." Drawing on the work of Deleuze (1993; Deleuze & Guattari, 1987), an object-interview is "an entangled conversational interview of objects and subjects, during which I mediated folds of objects, subjects, events, and a life" (Nordstrom, 2013, p. 243). Such an approach to interviewing de-centers the humanist subject (that is, the interviewee) from the focus of the interview so that her or his body-self is positioned among many "objects, subjects, and events [that] continuously fold together" into an interview that will "generate its own map of folding perspectives" (Nordstrom, 2013, p. 245). Nordstrom examined the entanglements of family historians (and their ancestors) with material objects such as documents, photographs, and family heirlooms. Likewise, Riley (2010) studied farmers and found that both photos and ledgers provided important points of departure for interviewees to discuss aspects of farming that were sparked by the interviewer's interest in these actants, the farmers' explanations of them, and subsequent questions. If I could revisit my previous ethnographic and interview study of dialysis care providers and their patients (Ellingson, 2008, 2011b, 2015), I would ask more questions about the dialysis machines, needles, dialyzers (filters used in the machines), bandages, and other supplies used in treatments in order to understand not only how the participants understood their wielding of these biomedical actants but also how the actants helped to constitute the material being of the health care providers and patients by engendering specific practices. Objects to be discussed in interviews may also include participant-created (visual) data (objects); these are discussed in the practices section of Chapter 6. While having the material actant available to be viewed, handled, or otherwise incorporated physically into an interview is ideal, photos or drawings of relevant actants might also be incorporated. When meeting in a participant's home, office, cubicle, employee lounge, or other participant-centered space, take the time to look for objects within the space that might spark sensory memories or details from participants and invite participants to comment on them.

Focus Groups

Focus groups are group interviews moderated by a facilitator who poses questions to the participants. This method of data collection enables researchers to ascertain how verbal and nonverbal cues are picked up from others in the group and shape the responses of individuals, which in turn shape subsequent responses. While researchers tend to focus analyses primarily on focus group participants' verbal responses to each other (i.e., speech content), ideas expressed are entangled with the embodied presence and nonverbal communication of the group interactions, including the facilitator (Belzile & Öberg, 2012). For example, one researcher who studied intimate domestic practices surrounding sustainability and water use (e.g., showering) argued that shared laughter (an embodied form of nonverbal communication) played an important part in enabling open discussion of personal practices (Browne, 2016). A study of menstruation and birth used the foundation of shared physical experiences unique to (biologically determined) women to generate meaningful and even transformational discussions of spirituality as grounded in natural bodily processes (Moloney, 2011). Discussions about sexuality in all-male, all-female, and mixed-sex focus groups varied significantly, demonstrating the entanglements of biological sex with individual gender performance and gendered group norms among peers and the (female) facilitator (Allen, 2005). Hence embodied participants—their presumed identity categories *and* their verbal and nonverbal communication within the focus group—intra-act with each other, cultural discourses, facilitators' questions and practices, and the particular place in which the focus group met to co-produce data with the researchers, who then construct analysis processes, study results, and shared representations of results. Whether researchers focus on the interactions of participants as meaningful processes or attend only to the verbal content, focus groups constitute embodied intra-action (Barad, 2007) and generate discourse that necessarily differs from that generated through one-on-one interviews. Focus groups can be used on their own to collect data or can be used to complement participant observation and individual interviews.

Debriefing Interview

Onwuegbuzie, Leech, and Collins (2008) recommended exploring researchers' "biases" via a reflective interview by a skilled interviewer who is uninvolved in the study on which a researcher is reflecting (see the suggested questions on p. 7). I agree with the potential benefits of such a process, although I would reframe this as a matter of reflecting on a researcher's standpoint, since bias connotes error or weakness that should be eliminated (or at least minimized) according to positivist standards. The "god trick" of an objective "view from nowhere" is impossible (Haraway, 1988), so being responsible to one's standpoint better reflects Onwuegbuzie et al.'s insightful suggestions for systematic reflection on the intersection of researcher standpoint (or biases) and particular details of a study's embodied processes and products. For example, the authors suggest asking, "How comfortable were you interacting with all of the participants? Which participants made you feel more/less comfortable?" (Onwuegbuzie et al., 2008, p. 7). The authors also note that their questions can be answered and recorded by the researcher without an interviewer, but they consider that approach less beneficial to the process of reflection. Ideally, externalizing the reflexive questioning process through embodied intra-action with another researcher offers a powerful way for researchers to think through the complexities of their own and participants' body-selves in the meaning making process.

Emplacement

The setting of the interview is entangled with the meanings co-constructed throughout the interaction. Place matters in the research interview. If possible, go to (some of) the everyday places of those being interviewed. Draw on all your senses as you explore meanings and accounts with your participants. The context of your bodies during an interview is important to the meanings constructed, not just incidental (Pink, 2009). Moreover, the focus on place can expand the scope of the interview to ask about unanticipated topics; "being *in* place and talking *about* (or *through*) place allowed the interview to 'bring in' other respondents and narratives" of farming, such as a farmer's wife, and interruptions became an impetus for explanations, clarifications, and other perspectives (Riley, 2010, p. 653; original emphasis).

6

TEXTUAL BODIES

Creating Embodied Data

Doing Legwork: Bodies of Data, Data of Bodies

Staff shrouded in white lab coats and rows of computerized equipment glowing in stark fluorescent light gave the dialysis treatment room a cold, mechanistic air. In contrast to this sterility, two dozen thin tubes of bright crimson, circulating blood bespoke the vulnerable bodies that reclined next to each machine. A vivid array of patients' colorful mittens, knit hats, sheets, blankets, pillows, and sleeping bags provided some cheer—the treatment room looked like a bizarre winter slumber party for senior citizens. Patients used outer wear and bed clothes to cope with the chilling effect of blood circulating outside their warm bodies in cool machines. Most of the twenty-five treatment stations were full, and I watched as staff members busily prepared the remaining stations for the next shift of patients. Technical aides rapidly stripped the tubing from empty machines, and patients who had completed treatment sat holding gauze on their blood access sites to encourage clotting. As I gazed around at the dialysis machine screens, green lights glowed, yellow and red lights blinked, and alarms beeped to alert patient care technicians of potential problems.

"Are you asleep?" I asked softly as I approached Mrs. Albright's chair. All dialysis patients sit in upholstered reclining chairs with their feet elevated during treatment to help maintain their blood pressure as their blood circulates through the filtering machines that approximate the function of kidneys for patients with end-stage renal disease. The middle-aged woman had her eyes closed, but I had learned that she often closed them without sleeping. Indeed, when I told Mrs. Albright that I hadn't talked to her during one previous research visit because I was afraid I would wake her, the patient had scolded me for assuming she'd rather sleep than chat.

At the sound of my voice, Mrs. Albright's eyes popped open, and her lushly painted, hot-pink lips curved into a smile. "Hello!" she said. "You came to see me!" She blinked as she readjusted to the bright fluorescent lights and scanned the busy room full of dialysis patients, patient care technicians, and other health care providers.

"Yes, I always have time to chat on Tuesdays," I replied, pulling a rolling stool over and sitting down. "It's nice to see you," I added.

Mrs. Albright reached down into her tote bag and said, "You too. I had my husband pick you some lemons, after our conversation last week." Smiling, she held out a plastic grocery bag with several lemons from her prolific tree.

"Oh, that was so nice of you," I said, touched that the older woman had remembered. I set the bag on the floor by her feet. "Thank you so much. We'll enjoy those. I'll have to make lemonade or lemon curd. So what have you been up to?"

"Well, you know, my husband finally took me shopping, and I ordered a new outfit for Mother's Day. We're going to see my older son and his family, you know, and we'll all have dinner that day, so of course I need something new to wear." Mrs. Albright smiled impishly.

I laughed. "Well, naturally! *Any* excuse for a new outfit is a good one." At the moment, Mrs. Albright looked cheerful and comfortable in one of her many expensively casual track suits, this one in a shade of pink that coordinated with her lipstick.

"We're going to The Hacienda. I can't wait," Mrs. Albright explained, proceeding to detail everything she expected to eat, not one item of which conformed to either the guidelines for managing her diabetes nor the dietary restrictions for dialysis. "Oh, there's Tim! Can you wait just a second, Laura?" Not waiting for my reply, Mrs. Albright waved to her patient care technician Tim, watching him intently and trying to catch his eye. "Tim!" she called.

Tim nodded at Mrs. Albright and held up his index finger. "Just a minute," he replied, turning to press a series of buttons on the touch screen of a dialysis machine.

"Oh good," said Mrs. Albright. "I want to talk to him about having more weight taken off. I want to go up to four kilograms."

I was puzzled and asked, "Your doctor wants you to have more fluid removed during your treatments?"

Mrs. Albright shook her head. "Well, she didn't say. But I would really like to be able to fit into that new outfit. I'd like it if they took more weight off."

I was unsure how to respond. I knew that Peter, the unit social worker, had carefully explained how dialysis works before Mrs. Albright started treatment (as he did with all patients and their families), and while I was no expert on the technical details of dialysis, even I knew that fluid removal was monitored quite precisely to ensure that patient's "dry weight" remained stable and only excess fluid was removed. "Um, they don't remove fat, you know. Just fluid and waste stuff, toxins in your blood, that kind of thing. Not your *real* weight, you know?" I hesitated, not wanting to offend.

My favorite patient stopped smiling at me. "They remove weight every time I come here, three times a week."

"Ah, I think they have to have orders from your doctor on how much fluid to take off," I said tentatively. Surely she didn't think dialysis functioned as a weight loss method, or that the amount removed each time was arbitrarily determined by the patient's wishes? Mrs. Albright frowned at me steadily and with a shrug, I retreated in cowardice. "Well, I don't know how all that works, you know," I said. "You should talk to Tim."

I was rewarded with a nod and small smile. "Yes. I will as soon as Tim gets over here. He's busy today, seems a little grumpy. Did I tell you Bob took me to the Rib Shack for dinner last night? We had baby back ribs and coleslaw and corn bread – it was delicious."

"Oh that sounds wonderful!" I said, thinking to myself, *and nonadherent with your treatment regimen.* "I made fish last night, with zucchini. Not as exciting as what you had."

"But better for you," Mrs. Albright said, shrugging. Unable to walk more than a few yards at a time, fatigued by the symptoms of kidney failure, suffering from intense side-effects of dialysis treatments, and left home alone much of the time while her husband worked, Mrs. Albright clearly felt eating was one of her few remaining pleasures. She had told me previously that she enjoyed coming to the dialysis unit since it got her out of the house and gave her a chance to talk to people. Her evident pleasure in my visits with her as I conducted an ethnographic study in the dialysis unit was gratifying but also a poignant reminder of how little social interaction the woman had. I did not want to ruin their chat or denigrate Mrs. Albright's need for pleasurable eating by pushing her to conform to the dietary guidelines, but I felt guilty and worried about the patient's health.

"And how's your kitty doing?" I asked, changing the topic.

"Oh, Ginger is fine, sleeps with me every night," replied Mrs. Albright, smiling broadly, warming to her topic. "Yesterday she—Oh, Tim," she said as he approached. "I want you to increase the amount of weight you remove today."

Tim looked at her blankly and turned to check the display screen of her dialysis machine. "I can't change the amount. Got to have a doctor's order."

"My doctor said it would be okay," said Mrs. Albright. I looked at her, wondering if that was true.

"I'll ask Gabriella," Tim replied, referring to the registered nurse who was the charge nurse that day. He wandered off without another word.

"Humph," said Mrs. Albright. I saw Tim approach Gabriella and watched as they talked.

Mrs. Albright and I continued to discuss our cats until Gabriella came over and stood by Mrs. Albright's feet.

"I called and left a message for Dr. Cho," said Gabriella. Mrs. Albright nodded. "But I can't increase the amount of fluid we remove without an order. What did she say last time you saw her?"

Mrs. Albright set her mouth in a stubborn line. "I told her I wanted more taken off, and she said it would be fine."

"Well, I checked the faxes, and we didn't receive anything from her," said Gabriella, smiling and speaking firmly. "I'll let you know when I hear back from her, but I can't change anything until then. Okay?"

Mrs. Albright sighed, clearly displeased. "All right," she said begrudgingly.

"I have to go," said Gabriella, hurrying away as she was paged over the intercom to take a phone call. Mrs. Albright turned back to me and began describing Ginger's latest kitty antics, and I smiled sheepishly at Gabriella as the nurse moved off. Out of Mrs. Albright's sight, Gabriella caught my eye, shook her head, and shrugged, then smiled at me before moving to the nurse's station.

An hour later I was stuffing my lab coat into the laundry bag in the dialysis unit's back hall as I did after each visit to my research site, when I saw Anne, the unit's registered dietician, walking down the hall with Peter, the social worker.

I joined them as Peter said, "Just had a nice chat with Mrs. Albright."

Anne rolled her eyes and asked, "You talked to her?"

"Yeah, we chat almost every week," he confirmed, nodding.

"Does she tell you about what she's eating? Her levels are *really* high; her labs [blood test results] are not good. And I tell her, 'You aren't watching what you eat.' Too much sodium, too much liquids." Anne shook her head, looking both annoyed and worried.

I flushed guiltily, remembering how quick I was to give up trying to persuade Mrs. Albright to be more reasonable in her eating. Peter nodded again. "Yes, she tells me about all her meals out—really rich stuff, lots of sodium, alcohol, whatever."

Anne sighed. "She is *clearly* noncompliant."

Peter looked thoughtful. "It helps to remember that Mrs. Albright is proud of being in charge and in control. She would rather be in control than be compliant." As a social worker, Peter spent time talking with each patient about how they coped personally with kidney failure and dialysis, and how their friends and families handled their condition.

Anne shrugged. "Yeah. Well, she is very definite about making her own choices. I tried to suggest that this is not the best for her, explained again how nutrition relates to kidney failure and dialysis treatment, but she didn't want to hear it. She was in the hospital again last month. Complications," said Anne. Shaking her head, she added, "It's not up to me—I do what I can to explain."

"She wants the technicians to remove more weight so she can be thinner," added Peter with a smile.

Anne laughed. "Oh yes, she wants to lose more weight. She lost some weight last time she was in the hospital." Anne shook her head in dismay and then shrugged. "There is only so much I can do."

"Yeah, that makes sense," Peter said kindly.

"I've to go run, need to pick up my kids. See you tomorrow," said Anne, nodding to me, then heading into her office.

"See you later," I said to both of them as I turned to leave, shivering in the cool air, my thoughts spinning as I reflected on my afternoon of fieldwork.

How and where are bodies written? I wondered.

What Do Data Do?

To begin, I consider what is meant by the term *data*. Generally, qualitative researchers (myself included) write about how we "collect" or "gather" data. We make audio recordings, digitally record conversations held in Internet chat rooms, participate in fieldwork experiences that we inscribe as fieldnotes, invite participant-produced artwork such as photographs, drawings, and collages, and find and record or document other texts and artifacts. Data is the term commonly used to label these empirical materials that form the corpus for analysis and the basis for knowledge claims made by researchers (except perhaps by those working primarily at the artistic end of the methodological continuum; see Chapter 3).

The term data comes with a constraining positivist legacy that connotes the discovery of sensory information that exists independently of researchers as "some thing that one gathers, hence is *a priori* and collectable" (Markham, 2013, sect. 1, ¶3; original emphasis). Moreover,

> [t]he term 'data' . . . functions as a powerful frame for discourse about knowledge—both where it comes from and how it is derived; privileges certain ways of knowing over others; and through its ambiguity, can foster a self-perpetuating sensibility that it is incontrovertible, something to question the meaning of, or the veracity of, but not the existence of.
>
> *(Markham, 2013, sect. 1, ¶2)*

Awareness among qualitative researchers that what we label data is not an objective or transparent account of reality is not entirely new, of course. In fact, most qualitative textbooks address the constructed nature of data. Anthropologist Geertz famously stated that

> what we call our data are really our own constructions of other people's constructions of what they and their compatriots are up to . . . Right down at the factual base, the hard rock, insofar as there is any, of the whole enterprise, we are already explicating: and worse, explicating explications.
>
> *(Geertz, 1973, p. 9)*

Critical methodologists point out that data—as part of all methodologies—are never neutral. Social scientists generally do not address the constructed nature of data and describe data in such a way as to deny its fabrication and the researchers' role in that process, through passive language (e.g., "the data were collected") and through specialized codes that establish the validity of data and remove traces of embodiment (e.g., criteria for rigor, theoretical saturation). Hence, looking to the postpositivism to redress the mind–body split is a doomed endeavor; as Audre Lorde (1984) explained, the "master's tools will never dismantle the master's house" (p. 112). Moreover, conventional methodological practices, especially if left uncontaminated with critical sensibilities, reflect modernist and masculine experiences, language, and sense making (DeVault, 1990). Critical and interpretive scholars have long resisted the framing of research "subjects" from whom the expert researcher extracts data. This terminology frames data as existing outside of the researcher and participants. Interpretive qualitative researchers understand epistemology, that is, what counts as knowledge or evidence, as influenced by social constructionist theory that suggests that data are co-constructed between researchers and participants (e.g., Charmaz, 2006) or postmodern and poststructuralist theorists who question the notion that research procedures can adequately capture the complexities of interaction and context (e.g., St. Pierre & Jackson, 2014).

Nonetheless, postpositivist, and most interpretive, postmodern, and even critical qualitative researchers still report "collecting data," despite our awareness—on some level—that the positivist axiom that data pre-exist researchers is countered by our theoretical commitments. The language of "data collection" is perpetuated by disciplinary and professional publishing standards and practices, as well as a certain pragmatic motivation to make "data collection" practices teachable to each new generation of qualitative researchers. Moreover, no one has launched a successful, practical alternative to this implicit framing of data as "out there" that does not simultaneously reject the notion of qualitative analysis. Researchers who embrace a "post-qualitative" mode problematize "data" as an assemblage of human and nonhuman objects and "data analysis" is rejected as inherently positivist, modernist, formulaic, and pointless. In such a framing, researchers "provoke discontinuation of data as we have come to know of it through post positivism, empiricism, text books, research training, and other grand narratives. . . . [and suggest] (un)knowing and (un)doing data" (Koro-Ljungberg & MacLure, 2013, p. 219) in favor of close readings of data assemblages through a theoretical lens (St. Pierre & Jackson, 2014).

While reading through theory is certainly one valuable mode of research, it, too, has its limitations. In particular, adoption of poststructuralist research practices to the exclusion of middle-ground qualitative data collection practices would limit audiences for research to an unhelpfully narrow contingent of those who have been trained by academic mentors to embrace and comprehend sophisticated critical theory. Unwilling to reject data analysis entirely and step completely outside of mainstream disciplinary research journals, researchers in the middle and between the middle and artistic end of the qualitative continuum—e.g., interpretive, feminist, or social construction-ist researchers—generally bracket the issue of what researchers do when we do data collection, not delving into these theoretical complexities when report-ing study results. Instead, these researchers embrace practices such as situating researchers' embodied identities in relation to our participants and topics and citing methodological treatises that provide theoretical and methodological context for our knowledge claims. I (and clearly many others) believe that our choices are the best we can do in the current academic climate, a form of "guerilla scholarship" as intervention into the important conversations among scholars in our disciplines (Ellingson, 2009a; Rawlins, 2007).

Even though we may not explicitly address the issue in our more conventional research reports, middle-ground and middle-to-art researchers agree that data are not a set of discrete objects picked up by a researcher as one would gather smooth pebbles by a river; nor do researchers and participants construct data together as they might use a set of building blocks to jointly fashion a tower. Instead, for the purposes of learning about doing embodiment, data can be thought of as multi-faceted texts, or *assemblages* (Denshire & Lee, 2013), that are produced through the intra-action (mutual constitution) of the researcher, participants (and other people in the setting), actants (objects), and cultural discourses within particu-lar places and times. We can welcome data that is not (only) collected but also (or alternatively) "wondered, eaten, walked, loved, listen to, written, enacted, versed, produced, pictured, charted, drawn, and lived" (Koro-Ljungberg & MacLure, 2013, p. 221). Moreover, researchers might do well to "[r]emember that the original meanin[g] of datum is 'somethin[g] given.' Data is a gift, so be thankful for it when it's given to you and treat it with respect" (Saldaña, 2014, p. 979). Data is a useful shorthand with unfortunate connotations that many qualitative researchers agree to use despite its limitations. Since my project in this book is to persuade researchers of the rich possibilities for incorporating embod-iment into their research processes and practices, I attend to the constitutive processes through which data come into being in ways that bring participants' and researchers' bodies into focus and position body-selves for reflection in order to capture the "the sensorimotor magnetism of the universe in question" (Wacquant, 2009, p. 123) and produce "stories in the flesh" (Warr, 2004, p. 586).

From here, I shift the question from "What *are* data?" to the poststructuralist-theory prompted, "What do data *do*?" Data intra-act in the world in a continual state of flux; data do not passively wait for analysis but exist within the world in an ongoing state of *becoming* data (Deleuze & Guattari, 1987). Daza and Huckaby (2014) explain that as researchers they reject the notion of data as static and instead consider it continually "becoming data" over time and through their ongoing experiences with the world. Like Childers (2014), myself (Ellingson, 2009a), and others, they embrace a wide range of methods and theories, gleefully crossing paradigmatic, methodological, and disciplinary boundaries and embracing the liminal spots between. Such a stance is not just a theoretical standpoint but a biochemical, body–mind process that produces the conditions for creativity . . . *Percolating data* refers to this active way of developing/analyzing phenomena on one hand, and letting go of how it is already developed, on the other. To percolate is both a process of filtering, permeating, and oozing-out; a generative process through which data sustains and gains energy and meaning.

(Daza & Huckaby, 2014, p. 803; original emphasis)

The percolating data metaphor evokes a wonderful sense of water filtering through an old fashioned coffee percolator, developing deep, rich, and complex flavors over time; the coffee does not stay suspended until the researcher has time to drink it but it changes in temperature, texture, and taste over time. Moreover, the data remain always inherently fluid and unstable: "Data will always exceed itself and evolve and transform as it intra-acts with other data and research assemblages" (Ringrose & Renold, 2014, p. 778). This fluid sense of data is generative and highly conducive to resisting the mind–body split to embrace a biochemical percolating of the body throughout all aspects of our research processes.

Researchers also are taught to construct data to focus on the meanings and words of humans, including other objects and the contextual space(s) under study as background material. Researchers should "acknowledge when looking at the creation of data within . . . this conventional interview . . . not only how it produces interesting knowledge . . . but also . . . how this is done in a particular way that makes the human(s) the center of attention" (Bodén, 2015, p. 196). Researchers generally choose to ask questions and make observations predicated on the perspective of humans as the crucial focus, thereby perpetuating the construction of data with the same humanist focus. Yet other possibilities exist to illuminate complex networks of actants, humans, and contexts as they intra-act. For example, the phenomenon of global poker positions humans in a complex technological web or network but not at the center of inquiry: "The interaction of these technologies and their human participants constantly changes how the game is reported, played or watched" (Farnsworth & Austrin, 2010, p. 1121). In this research story, the technologies of social media, television, mobile devices, and the Internet construct the humans as much as the humans construct and utilize the technology. Discourses of gambling, card playing, professional and amateur poker competition, online gaming, masculinities, and global capitalism (among others) are woven throughout the people and technologies of global poker. These authors include humans but do not privilege them as the only agents in the poker phenomenon.

Clearly, data are not passive objects but constructed, becoming actants in themselves. Next I discuss a variety of approaches to constructing data that does embodiment with humans, actants, and discourses in a number of modes, including fieldnotes, recordings, transcripts, and participant-created data.

Fieldnotes: The Language of Embodied Data

The goal of ethnography is to render bodily experience and practice into language on the page (and screen, canvas, stage, and so on): "Ethnography is nothing until inscribed: sensory experience becomes text" (Fine, 1993, p. 288). Not just getting down words and ideas, but making a visceral, messy, sensorial, richly evocative text poses challenges. Fieldnotes are not (nor should be they be) a record of what ethnographers thought while we observed, interviewed, or otherwise participated in the moments and contexts about which we make sense (Emerson, Fretz, & Shaw, 2011). Instead think of intermingled sensory experiences transformed into data that remain fluid and evolving. The process of participating in *becoming* fieldnotes arises in response to questions such as these:

> How to go from the guts to the intellect, from the comprehension of the flesh to the knowledge of the text? . . . [How] to restitute the carnal dimension of ordinary existence and the bodily anchoring of the practical knowledge constitutive of . . . every practice, even the least "bodily" in appearance[?]
>
> *(Wacquant, 2009, p. 122)*

Inscribing fieldnotes is messy and often tedious. Embracing a multisensorial approach that highlights the open-endedness of data and interpretations rather than fixing them often proves a significant challenge. Guts and intellect come together when researchers inscribe embodiment; the following are three considerations that offer a helpful framework for doing embodiment in fieldnotes.

First, writing fieldnotes is not exclusively a mental process but a material one as well. Writer Natalie Goldberg suggest that writers understand writing "as an embodied practice" in which one engages not only cognitively "but with your whole body—your heart and gut and arms. . . . You are physically engaged with the pen, and your hand, connected to your arm, is pouring out the record of your senses" (Goldberg, 1986, pp. 37, 50). This sensuous record must be drawn from ethnographers' body-selves, not merely (what we think of as) our brains. Ethnographers can remind ourselves to write *through* our bodies as we construct fieldnotes in order to promote a mind-body stance conducive to sensing and articulating the connections between our bodies writing fieldnotes and our bodies as lived with our participants' bodies.

Second, ethnographers should construct fieldnotes with the explicit recognition that meaning is not fully contained in language or transparently made available through language. Language is symbolic and indeterminate; its meanings are multiple, unfixed, and ambiguous. Interpretive and critical researchers can "assume that language is always already contaminated with meaning, exploding with meaning deferred. In these approaches, language cannot and never has been brute" (St. Pierre & Jackson, 2014, p. 716). Researchers cannot utilize language to construct (or analyze or represent) data as though its meaning were given, straightforward, or singular. Cultural baggage—implications, allusions, connotations, allegiances—weighs down language with ideas, and those ideas are rooted in bodies.

> The world of signs and meanings is *made*. . . . [A]nd in its making it gets anchored in what *appeals* to the senses . . . Meaning is body centered, anchored *in* the senses, and frequently *about* body conditions—a measure of how we are at any given moment, a platform for interpreting the "stuff" of our lives.
>
> (*Brady, 2004, p. 624; original emphases*)

Researchers make that meaning in our body-selves. When we write fieldnotes, we can *do* our body-selves in part by keeping Brady's idea close to our hearts—that is, that all symbols (i.e., language) appropriate sensory experience within them. Yes, language can be used to distance ourselves from our data (and academic traditions encourage it), but language also can be used to infuse data with senses and bodies, our own and others. The links among senses, words, and ideas are always in the process of becoming, and ethnographers can deliberately join in the process when we inscribe fieldnotes *through* our bodies.

Third, expressing richly sensorial experience relies on imaginative and figurative language in construction of fieldnotes (not just in representation of results). "[T]he English language is full of sensory and even haptic metaphors . . . Poetry is full of sensory conjunctions achieved through simile and metaphor, and allows an alternative pathway to the sensory 'reporting back'" (Paterson, 2009, p. 785). Poetic or otherwise figurative language is an important resource for inscribing embodied experiences, and it needs to be chosen with care. Fieldnotes are "only" data, and typically are thought of as private and preliminary. However, they are the primary sensory record generated by fieldwork, and the richer and more evocative the sensory language in fieldnotes, the more useful they may be for later sparking imaginative analyses and reflection. Sensory language in representation is discussed further in the practices for fieldnotes offered later in this chapter and in the discussion of representation in Chapter 8.

Recording Bodies

Qualitative researchers generally are socialized to accept transcriptions as authentic records of interviews and not to spend much effort contemplating their construction processes. Critical and interpretive scholars acknowledge the constructed nature of transcripts, if not fully embracing their material effects. Yet even before the construction of transcripts, researchers construct recordings that also are material artifacts, and the ways in which recordings are made have powerful effects on the (dis)embodiment of data. In this section, I review critiques of recording and transcription practices as they relate to doing embodiment through data.

Recording

Creating an audio recording is not a neutral or inconsequential act, yet qualitative researchers tend to treat them as though they were, manipulating and drawing on them in our constructions of participants' meaning. I suspect that interpretive researchers, even some postpositivists, and all researchers who draw on critical theories would acknowledge that recordings are constructed and have definite limitations. I make audio recordings as a regular research practice, and occasionally use video recordings as well, and I find them quite useful. In the same way that finding a practical alternative to the concept of "data collection" generally has eluded qualitative researchers who want to publish in mainstream disciplinary journals, we can note the limitations of recordings and yet use them anyway as a rich resource. We can endeavor to complexify our treatment of these resources rather than eliminating them; even the "post-qualitative" or "post analysis" researchers still use audio and visual recordings at times as the basis for their close, theoretical readings of data (Childers, 2014). Several aspects of recordings relate to doing embodiment: their emplaced construction through intra-action of people and actants, the types of technology used, researchers' interpretation/prioritization of sounds and noises in the recordings, and acknowledgment that the recordings both are missing things present during the recording period *and* have things added to them that were not in the original interaction.

First, recording devices intra-act with people and other objects as they record (Barad, 2007). The presence of recording devices (i.e., audio recorders, video recorders, cameras) is neither innocent nor inconsequential but "are a taken-for-granted material-discursive practice" (Nordstrom, 2015, p. 389) in which, as researchers, "we (recording device, cultural conceptions, participant, bodies, and the physical space of the object-interview) constitute each other as well as produce an articulation of the dynamic world" (Nordstrom, 2015, p. 394). Recording is a consequence of many technical choices and of the available equipment (Modaff & Modaff, 2000). Digital recordings are standard now, but for many years the material oddities of cassette tapes and reel-to-reel tape were highly relevant. Analog recording challenges have been replaced by the difficulties of operating digital technologies, the ease of clicking in the wrong place on a touch screen, the crashing of computers, and the intermittent failure of wireless Internet access.

Researchers now face many choices among customizable settings for digital recorders, and microphone type, quality, positioning, and functioning remains another set of vital factors. Many practical matters and technical details such as signal variation, quality of recording, tape properties [or digital file properties] also matter in the production of the material recording (Modaff & Modaff, 2000). Background noise also is consequential and requires interpretation; "sound recordings are always liable to register a democracy of noise; but this can be fixed too, in the edit. Speech can be amplified in playback, background sounds suppressed or filtered" (Hall et al., 2008, p. 1025). A recording may also change the level of different sounds, highlight ambient noises that participants easily ignored or did not notice, or, alternatively, completely fail to pick up relevant ambient noises and subtle features of the conversation, depending on the type, location, number, and quality of microphones used to make the recording. Moreover, as any music lover knows, the mixing of sounds is a complex process and has a dramatic effect on the effect achieved by the final recording. Further, it is impossible to record a conversation at the precise levels and balance at which it was experienced by each participant in the space, resulting in at least some distortion of "what really happened" even with good quality, properly functioning equipment.

Recordings, then, do not collect pre-existing, or even co-constructed, reality, but intra-act to produce a slice of a dynamic world that exceeds researchers' questions and intentions. This does not mean that recordings are bad; while the creation of a recording is not a neutral act, it is also not an invalid, unethical, or useless act. However, recordings have historically been treated as transparent access to interviews (or other interactions) that contain raw, static data that wait for excavation and organization by researchers. This impression is fostered by the presumed static nature of the recording that has "captured" and fixed meanings in a particular time and place, an impression or implicit claim which I now seek to problematize. Think for example, of archived recordings of sexuality researcher Alfred Kinsey's classic interviews investigating individuals' sexual practices. Those recordings, made on simple analog recorders, no longer "contain" the same data. As they intra-act with the embodied people, actants, and discourses of sexuality in the twenty-first century, these recordings participate in very different "meanings" that lead to different "results" and material implications than they did when originally produced. Moreover, those recordings would not be understood the same way by all people in all cultures today either, but would intra-act differently in various cultural and institutional contexts. In the same way, any recordings researchers make now will contain shifting, unstable meanings that will vary across time, place, researchers, institutions, and cultures.

Recordings tend to be treated more as realist objects than are transcripts in part because recordings happened earlier in time; they were created closer to when the events occurred, and thus recordings appear more real, despite the many details they leave out and the crucial factor they add (Ashmore & Reed, 2000). Every time researchers listen to recordings, they change their understanding of their content.

> The interpretative and productive act of *listening* changes the Tape's status from an unknown to a known, from an object that is radically unstable to one which is relatively fixed. *Listening* polices the Tape. The "rules for hearing" distilled from this process are articulated in and as the Transcript.
>
> *(Ashmore & Reed, 2000, ¶34, original emphasis)*

Each time the recording is listened to and interpreted into transcript, the recording is changed, not just experienced. "To *question* what is heard or seen is to immediately render the object once more as produced, crafted, worked-on-and-worked-up. The evidential adequacy of any object is momentary, contingent and fragile" (Ashmore & Reed, 2000, ¶41, original emphasis). Qualitative researchers commonly treat recordings as though they were stable and contained meanings, bracketing their dynamic nature. The "recording devices and researchers attempt to grasp always already contested meanings that are part of and produced by dynamic intra-actions" (Nordstrom, 2015, p. 395). Nordstrom urged researchers to engage with recording apparatuses in ways that help make them accountable for how these apparatuses intra-act with other humans, nonhumans, and discourses, rather than treating them as neutral accounts of reality.

Further, the very act of recording a moment changes it; the act produces a different object. No immediacy exists in a recording—it offers a different "now." Interviews happen in a specific moment, that is, they are ephemeral and can be experienced only transiently by those present. Hence an interview "can be, in our terms, *seen* and *heard*. But it cannot be *read* nor *listened* to" (Ashmore & Reed, 2000, ¶25, original emphasis). So the interview is transformed from a past, lived, and remembered event to a recorded version that lacks much of the complexity and contextual cues. It can be re-experienced in part, but the interaction has been removed from its time and context and become fundamentally disembodied (this is true in video recording as well, in that the video recording is a digital object with *images of bodies* as they were then, not bodies themselves now). Awareness of this difference is important because researchers should avoid romanticizing the recording as authentic, particularly when we claim that a participant or informant meant something or intended to communicate a particular meaning. Absent the immediacy of the interview event, we are left with a recording of "then" and of "them," that is, researchers' past selves and participants' past selves, a slice of whom becomes frozen at a time that no longer exists, as body-selves that they cannot still inhabit—and this is true an hour, day, month, year, or decades after the interview took place.

Of course, important nonverbal communication cues will be missing from a recording, along with all sorts of contextual aspects of the setting. Sometimes these are not relevant to a given analysis, but shades of meaning are embodied in eye contact, facial expressions, gestures, arrangement of people's bodies within a space, and the like. Even video recording loses room temperature, feeling of chairs or other furniture, odors in the air, gestures not completely visible from the camera angle, and so on. Moreover, the recording also "introduces novelties *that were not there in the "original"*; chief among these being "replayability" (Ashmore & Reed, 2000, ¶26, original emphasis). That is, the event is not only fixed but is (barring technical failures) endlessly replayable, a capacity that living in ordinary time lacks. Hence, when the interview is replayed, it is fundamentally not the same; its character has been altered when it moved from an ephemeral moment to a replayable object. The interview conversation has new ways of being experienced that are revealed through subsequent listenings, as nuanced or unclear moments become clarified—and therefore experienced in different ways than they were at the time—through repeated scrutiny and reflection. Again, I do not advocate abandonment of recordings, only that researchers reflect on their construction processes and the ways in which bodies are translated and interpreted in replayable recordings of past moments.

Transcription

A transcript of a research interview or other recorded data is "a clichéd representation of the interactions that occur between two people [or more] at one particular moment in time" (Honan, 2014, p. 14). Warr (2004) offers an apt metaphor for the limitations of transcriptions, which is "akin to knowing a city by studying it only as a map. It is a poor substitute for actually getting out into the field" (p. 584). Far from a neutral act of transferring of words from a recording to a page—as understood in positivist and postpositivist paradigms—transcription is an act of translation and interpretation between two vastly different modes of communication (Mishler, 1991; Warr, 2004). Like the recordings just discussed, "[t]ranscripts produced by the intra-actions during and after interviews, then, are contested texts that constitute, not represent, realities and meanings" (Nordstrom, 2015, p. 395).

Stripped of most nonverbal interaction cues and with little or no descriptions of the participants or setting, standard transcriptions of research interviews focus on accurately representing spoken language, placing rhetorical and visual emphasis on the verbal content while ignoring the differences between oral and written speech. The rhetorical choices inherent in transcription typically erase bodies completely, leaving only words and sometimes series of cryptic marks that reflect conversation analytic codes intelligible only to researchers trained in the techniques and not at all equivalent to the nonverbal cues for which they stand (Del Busso, 2007; Mishler, 1991). Conversational analysis includes in transcripts a variety of coded nonverbal cues such as drawing in breath, vocal stress on syllables, rate of speech, and so on, all of which are bodily enacted by lungs, esophagus, lips, teeth, and tongue. This form of transcription is very time-consuming, and, while quite valuable, does not fully re-embody transcripts either (Hawkins, in press). Mishler (1991) persuasively argued that moving from oral to written speech involves not accurate dictation and notation but instead a form of translation. Just as ideas moving from one language to another may make it impossible to express many ideas accurately in their direct equivalent in the other language, writing down oral speech renders it a different entity altogether. People do not process oral speech the same way we read it; we encounter and make sense of language in vastly different ways depending upon how bodies take in the cues. Mishler critiques accuracy as a naive standard for evaluating transcription; retaining the actual *words* in written form does not convey the *truth* of what happened in the (oral) moment.

Davidson (2009) invoked Ochs (1979), whose "central claim that 'transcription is a selective process reflecting theoretical goals and definitions' (p. 44) still stands as unrefuted" (p. 36). Researchers naturalize the process of transcription as neutral when it is better understood as interpretive and involves making many choices (Davidson, 2009; Duranti, 2006). Mishler (1991) argues that the transcription is part of data analysis, since decisions about how to represent the interactions on the page already frame future study results by structuring the discourse on the page in ways that reinforce assumptions researchers bring to the question, data, method of analysis, and anticipated results. Transcription reflects theoretical and epistemological assumptions, active choices by researchers of which transcription practices or notation systems to use, selectivity (what to include and omit), instructions given to transcriptionists (if other than the researchers), and resolutions of (dis)agreements between transcribers (researchers) when listening for components of talk to be represented (Davidson, 2009), all of which point to embodied aspects of speech and relational (turn-taking) aspects of conversation between embodied beings. Moreover, the analytical decisions do not just inform the transcripts; "the author's analysis . . . instructs the reader in what to hear on the tape and what to see in the transcript" (Ashmore & Reed, 2000, ¶44) (see also Green, Franquiz, & Dixon, 1997). Moreover, "reification of the transcript as the primary artifact of the interview extends from a logic formation that privileges a voiceless-voice, one that draws from an all-too-easy separation of the discursive from the material" (Kuntz & Presnall, 2012, p. 733). Such accounts obviously present only a very partial representation of "what happened," and of course these disembodied accounts do not favor all participants equally; "[t]he power of an embodied voice, which can deliver a sense of struggle, despair, or resilience, is greatly watered down when it is transcribed into mere words on a page" (Warr, 2004, p. 581). This watering down continues when the transcript leads to further analysis, then yet again when the analysis leads to representation in a journal article featuring small, decontextualized excerpts of participants' talk.

Researchers also must make decisions about "cleaning up" transcripts to make quotes more comprehendible and credible. This process is risky, however. When researchers clean up the speech in a transcript, we lose vital clues that hesitations, pauses, and struggles to rephrase an idea may indicate (DeVault, 1990); such verbal disfluencies often relate to gender, race, class, and other marginalized experiences which the dominant language is not well suited to express

(Reinharz, 1992). Retaining "inelegant features" of participants' talk shows their distinctive perspectives and also may signal embodied emotion, which is often ignored in analysis and rendered to seem like rational talk (DeVault, 1990, p. 109). Participants' speech is often cleaned up so that the written account looks and reads as more credible and is less likely to cast participants in a poor light based on group stereotypes.

Moreover, other issues pertaining to "cleaning up" can arise. For example, when transcribing interviews of participants for whom English was not their first language, one researcher explained:

> I chose to correct the frequently misused personal pronoun (he and she) and to elide long stammering associated with struggling to find the correct English word. I left the hesitations and struggles in the text when these reflect emotional rather than grammatical struggles. Differentiating between these two was surprisingly easy to discern in the context of conversation, although other decisions regarding interpretation were less clear.
>
> *(Swartz, 2011, p. 61)*

The complexities of transcription make it impossible to anticipate all the taken-for-granted ways in which researchers make decisions about transcripts. Decisions are unavoidable; the challenge is to be response-able to what we choose, that is, to be able to respond critically and reflect on the implications of our choices for embodied sense making, including in largely taken-for-granted methodological processes, such as recording and transcription.

Participant (Co)Created Data

Participant-created data endeavors to embrace embodiment in participants' daily lives more fully than researchers typically do when they inscribe data. Of course, all research depends upon participants' participation but participatory approaches share power and control with participants more equally than do traditional data collection practices. Greiner (2012) offers a continuum of participation for community members from respondent, through respondent and analyst, to (co)researcher. Sharing power and control enables bodies to move more freely, less constrained by embodied power of researchers with their academic credentials, and emphasizes the value of participants' perspectives and knowledge that is grounded in their daily lives (Horowitz, Robinson, & Seifer, 2009; LeGreco, Leonard, & Ferrier, 2012; Thorp, 2006; Wang, 1999). And yet sharing control also can prove problematic to actualize in practice (Wagner, Ellingson, & Kunkel, 2016), and social change can be difficult to actualize, despite good intentions (Sanon, Evans-Agnew, & Boutain, 2014).

Moreover, participatory approaches often blend with arts-based research practices (ABR) (Cahnmann-Taylor & Siegesmund, 2013; Coemans, Wang,

Leysen, & Hannes, 2015; Faulkner, 2016; Leavy, 2015) which use artistic formats or mediums (or *shapes*, in Leavy's terminology) to express themes or key insights in research results. Arts-based research techniques "represent spaces of agency and struggle for conscious, creative and critical expression" (Singhal & Rattine-Flaherty, 2006, p. 315). ABR accesses other types of knowing and being and creates embodied data because ABR is often visual, and sometimes musical (Ledger & McCaffrey, 2015). ABR complements written and spoken texts, rather than replacing them. "Textocentrism – not texts – is the problem," explained passionate advocates for participatory ABR with marginalized people (Singhal & Rattine-Flaherty, 2006, p. 315). ABR can use written texts as well, inviting poetry from participants, which has rhythm and cadence that is felt viscerally (Faulkner, 2009). Strategies for inviting participants to create creative, visual data through ABR are discussed below and in Chapter 8.

Practices for Inscribing Bodies in Data

Fieldnotes with a Single Sense

I advocate in this book that qualitative researchers think of senses as integrated and holistic. Nonetheless, as an exercise and a source of potentially contrasting or complementary data for your ethnography and/or interviews, try focusing on one or two senses at a time. For example, one ethnographer sat in a hallway of an in-patient hospice, without being able to see much from her position, and what she noticed were sounds and smells "such as beeping machines, patients sobbing, and meals being wheeled into patients' rooms, and the smells of antiseptic, drugs, and food, which permeated the ward" that gave her valuable insights on what patients experienced from inside their own rooms (Wray et al., 2007, p. 1396). Try spending time in the field for your study discerning and describing each of the following in turn for an hour each: smells, textures and temperatures of objects, movements of people and actants, and background/ambient noise (any sound except participants' speech). It may be helpful to separate senses out in order to turn conscious attention to different types of sensory experience and go deep for details.

Storytelling

Even if you have no intention of publishing ethnographic narratives or auto-ethnographic narratives (you never know!), write some as an exercise in placing your participants (and yourself) back in your bodies through multisenso-rial descriptions. Choose an especially emotional or provocative interview or an everyday moment from your fieldwork and write a story of "what happened," trying to make it come alive with sensory detail of embodied people. Narrative knowing features characters, emotion, action, and plot, in contrast to analytical knowing, which emphasizes order, clarity, and facts (Ellis & Bochner, 2000). The goal of writing autoethnography is to illuminate an experience in evocative prose that invites readers to reflect on the ethical, cultural, and personal meanings of the narrative (Ellis, 2004). Berger and Ellis (2002) provide guidelines on writing both stories in which the writer is the main character and those in which the writer writes personally about the intersections of his/her stories with others'. This process involves sitting down with your ethnographic fieldnotes, personal journals, notes made during introspection, and/or transcripts of interviews, and using those materials—along with your memories—to construct a narrative. When I wrote an ethnographic "day-in-the-life" story of the geriatric oncology team that I studied, I began by reading through my fieldnotes, interview transcripts, and personal journals repeatedly, noting which interactions were interesting to me. I asked myself a series of questions as I read the notes, marking the interactions that might make a good story. Narratives show rather than tell, and writing a story provides you with an opportunity to show bodies at work, at rest, interacting, or being silent, and so on, making them a rich form of embodied sense making. Such stories can be used as the narrative equivalent of analytic memos that focus attention on and interpret certain aspects of your data that represent a theme or important point. Stories can also lead to or become part of research representations reflecting the artistic end of the qualitative continuum (see Chapter 2 for the qualitative continuum, Chapter 8 for more discussion of embodied representation).

Express Entanglements without Centering the Researcher

Pointing to the way in which English frames the subject of a sentence as preceding the predicate (or the noun coming before the verb), Nordstrom (2015) argues that this humanist language convention influences researchers' thinking to literally put ourselves first in our sentences and thinking, and hence at the center of our analyses. She urged qualitative researchers to "unlear[n] the habit of 'I'" and "learn different ways to express entanglements that throw into radical doubt language" (Nordstrom, 2015, p. 397). Experiment with your use of language in fieldnotes or analytic memos to render yourself something other than an authoritative "I" who sees and does and speaks. For example, Nordstrom explained that in one project she experimented with speech using "gerunds as nouns [which] created a contingent and shifting space of entangling humans, nonhumans, nature, culture, and so on without centering those movements with the use of 'I'" (Nordstrom, 2015, p. 397).

Gerunds are the noun form of a verb. For example, to walk is a verb (e.g., I walked to work); walking, when referred to as an activity or practice, is a noun and therefore a gerund (e.g., I enjoy walking when I am on vacation). Following Nordstrom's lead, I reimagined my fieldnote, "I spoke with Shayna, a nurse, as soon as I entered the clinic," to incorporate gerunds instead of myself and my nurse-participant: "Researching Communication spoke with Caring for Patients, as soon as Researching entered the clinic." This admittedly clumsy observation nonetheless provides a tiny crack in my humanistic centering of me and of the nurse-participant by centering the embodied processes that are happening—researching communication and caring for patients—that include us but also involve far more than just us in a complex network of people, practices, laws, disciplines, objects, and discourses. In this way, my identity and that of the nurse are subsumed under the embodied practices in which we were engaged. Of course, Shayna and I were also engaged in other practices (gerunds) that we shared, at least to some degree, including Performing Femininity, Behaving with Professionalism, and Maintaining Patient Confidentiality, among others. And the patient for whom Shayna was caring was likewise enmeshed in performances of her body-self in a network that included but extended outside of the clinic at which I was conducting an ethnography. Similarly, you might rewrite sentences of fieldnotes so that you are not the subject of the sentence. For example, "I walked into the clinic and noticed Mr. Harrison had already been put on his [dialysis] machine and was napping in his usual station on the right side of the room," could be transformed into: "A dialysis machine filtered Mr. Harrison's blood and a recliner supported his prone body as he napped in his usual station on the right side of the room, not noticing me as I walked into the clinic." Changing the wording in these—and other—ways can disrupt the taken-for-grantedness of human beings as the starting point and ending point of all analyses and opens up opportunities for sensing the entanglements of humans, actants, and discourses in novel ways.

Problematize Punctuation in Transcription

Punctuation is another largely unaddressed aspect of transcription that relies on the judgment of the researcher to determine how oral sentences should be punctuated in written form. Researchers intuit where sentences stop and start; often I have disagreed with how a transcriptionist divided up a sentence into two separate sentences when I would have left them joined with a semi-colon or comma. Sentence fragments, frequent in speech, also may prove challenging. Scott (2015) addressed this challenge in a way that reinforced the embodied nature of spoken language by omitting punctuation that grammatically structures sentences, such as commas and periods, and inserted a line break each time the interviewee paused

> so that the texts resemble verse, and the reader can access storytellers' unique speech patterns by reading the texts aloud (Peterson and Langellier, 1997). I also preserve each utterance without edits (leaving ums, ahs, stutters, etc.) and include my interpretations of narrators' emotions in order to draw attention to each individual's embodied performance of self in creating the narrative.
> *(Scott, 2015, pp. 229–230)*

This approach to incorporating the embodied performance of self by participants may prove useful, especially for those whose participants shared stories with unique rhythms that might be expressed in this structure. This approach may also benefit researchers even if only carried out on one or two transcripts as an exercise to compare to transcripts that were transcribed in a more conventional manner. You might also use this strategy as a jumping off point for reimagining oral punctuation and timing of speech in another creative manner. For example, what if you did not begin a new line with each speaker in turn but instead let speech flow together in paragraph form, beginning a new line each time a new topic was addressed? Viewing the same speech arranged on a page in different ways may help researchers to understand participants' words and selves from other perspectives.

Say Your Data

Transcripts are typically labor-intensive and require many hours of typing. Hawkins (in press) did not type her transcriptions. Instead, she listened to and repeated each line of her interview recording into voice recognition software that produced a written transcript. This process of repeating her participants words was generative in "bringing the researcher into a space of sitting with the data" (n.p.). She did not "fix" or otherwise edit oral features of participants' speech. Sitting with data has the potential for a visceral re-experiencing of herself and her participant in the interview context. Perhaps speaking the participants' word enables a more sensory experience of them than typing affords. Likewise, Perrier and Kirkby (2013), drawing on Brooks (2010), discussed use of voice recognition software as "embodied transcription" (p. 104). They argue that this method of transcription fostered a deeper, more nuanced understanding of participants' narrative and closer attention to nonverbal cues such as sighs and pauses. "Speaking words resonated with us; by attending to how we felt speaking from our participants' perspectives and our reactions to their thoughts and stories, we could better understand our position with respect to our data" (p. 105).

As with other creative strategies for doing embodiment, this spoken transcription process may work well as an meaningful exercise even if you do not choose to employ it for all of your project's transcription. Try saying a passage from a recording into voice recognition software (if you do not want to invest in a voice recognition software program, you can use a smart phone or other tablet, most of which have the built-in capacity to translate talk to type), and then type it in your usual way as well and reflect on the differences in your understanding of the words and your perception of their tone, rate, volume, and other embodied nonverbal contributors to meaning. This technique also points to ableist assumptions; I wrote the earlier discussion of transcription and embodiment as though all researchers type transcripts using standard keyboards and reinscribed typing as normative, when for many people using voice recognition software is their primary mode of computer use (Goodley & Runswick-Cole, 2013; McRuer, 2004).

Photovoice

Photovoice is a democratic method that widens the circle of who can express themselves as a research participant, including vulnerable populations such as people who are homeless, living in poverty, or diagnosed with a life-threatening disease such as cancer—all of which are deeply embodied phenomena (Novak, 2010; Singhal, Harter, Chitnis, & Sharma, 2007). Photovoice involves taking photos of everyday life that often feature people, commonly used or valued objects, and elements of participants' environments, enabling participants to express relationships and for researchers to gain insights into intra-action within participants' worlds.

Moreover, the use of cameras involves moving about while taking pictures, which adds a kinesthetic element to participants' sense making and perhaps brings them into contact with a larger variety of sensory experiences as they move from place to place or within one place. Photovoice and other methods involving "participant generated images" may be empowering to participants and may help bring in concepts and bodily experiences that would not come up in an interview or are difficult to explain (Balomenou & Garrod, 2016). Wang (1999) provided a

FIGURE 6.1 The Sheer Beauty of Nature.
Participant photo by Daniel, Voicing Survivorship: Stories of Long-Term Cancer Survivorship.

Courtesy of Daniel Russell (photographer).

useful, step-by-step explanation of the method and advocates its use for exploring women's health issues. An excellent photovoice manual is available free as a PDF file for download from photovoice.org.

Participatory Sketching

As with photovoice, participatory drawing or sketching provides participants "an expressive channel to voice their inner stories, as well as an active and empowering stake in the research study. Furthermore, because of its playful nature and its lack of dependence on linguistic proficiency" it can be used with those who may be reluctant or less proficient with verbal skills, including children (Literat, 2013, p. 12). Drawings can also act as a gateway to increased awareness of how senses are intertwined with experiences (Harris & Guillemin, 2012; Tracy & Malvini Redden, 2016). In addition to participatory drawing, Greiner (2012) highlights two forms of participatory sketching that have been particularly fruitful for health communication research in global contexts but which have utility for a wide variety of contexts, including: *network maps* which involve drawing a participant at the center of a page and asking him or her to draw lines to (figure or symbols of) all the people they work with on a given program or issue, enabling researchers to inquire about these interrelationships; and *life maps* which involve charting series of important life moments so that they can be discussed and reflected upon (see also "time lining" method, Sheridan, Chamberlain, & Dupuis, 2011). As with photovoice, drawing/sketching is good for working with vulnerable populations or addressing sensitive topics, and this method may help to share power with participants (Liamputtong, 2006). One study used participatory sketching as a tool to address students' public speaking anxiety (Rattine-Flaherty, 2014), another invited U.S. servicemen and servicewomen involved in the Iraqi war and their loved ones to draw maps that "chart land marks" for them in their experiences pre-, during, and post-deployment (Thompson, 2010), and still another invited sketching in a study of community responses to diabetes among under-served people in Appalachia (Wright, 2009).

Participant Journals

Researchers may invite participants to keep journals or diaries around a specific type of task, behavior, or topic (Beckers, van der Voordt, & Dewulf, 2016). Journals are often written (Waskul & Vannini, 2008) but may also be constructed in video or audio format. One study used video diaries of university basketball team members to explore themes of everyday life, identity, and the body; to further the embodied presentation of their analysis, the researchers included links to clips from the video diaries that are hosted on the journal's website (Cherrington & Watson, 2010). Creative examples of participant journaling abound: video diaries (Bates, 2013), audio diaries (Bernays, Rhodes, & Terzic, 2014), and email diaries (Jones & Woolley, 2015).

Create Multimedia Transcripts

Another possibility for doing embodiment in transcription is to integrate images and video or audio clips into your transcripts and fieldnotes to make them into multimedia representations. Using photos from photovoice projects, participant drawings or maps that you have scanned (to generate a digital image), or clips from recordings, visually connect images and sounds to the words of a transcript to facilitate multisensorial knowing through the intra-action of the visual, written, and auditory. In a study of family histories, one researcher reported that she chose to add photos of relevant objects to her transcripts to reference what participants talked about: "The objects, participants, discourse, culture, and I intra-acted together during the interviews, and I could not tease those intra-actions apart in the transcripts" (Nordstrom, 2015, p. 395). New possibilities for incorporating images and linked audio into transcriptions are possible with advances in computer technology (Davidson, 2009). Like conventional transcripts, multimedia transcripts are partial and constructed representations; however, they do create visual connections between words and images and sounds that enhance attention to the voices of particular bodies within the transcripts.

Care for the Bodies That Write

Creating data through recordings, fieldnotes, transcripts, and participatory images requires extensive writing to organize, create, collect, document, and analyze. As discussed earlier, writing is embodied work, not merely a cognitive exercise. Gale (2010) eloquently and playfully explains:

> Writing, then, is an embodied experience, a social experience. The inspiration might not be divine, but the product of a set of suitable circumstances, environmental, personal, emotional. For me, the most important thing is "having my own head space" so that I can concentrate, which means minimizing the negative influences of my body-in-the-environment, so I need to make sure that I have had enough sleep, no hangover, a quiet environment (a "room of one's own"; Woolf, 1989), and that [I] am in the midst of no emotional crises. Easy!
>
> *(Gale, 2010, p. 219)*

I assume that Gale's exclamation of "easy!" is meant in a joking or ironic manner; certainly I read it as such, given that my own intra-actions as a researcher (and those of all researchers I know personally) preclude the complete avoidance of the "negative influences" on my body-self. On the other hand, if we interpret the standard as an ideal that cannot not reached but should be aimed for, then I concur with its importance. Self-care is just as important for writing as it is for any other activity in which we want our body-selves to engage (see discussion of self-care in Chapter 2).

7

ANALYZING BODIES

Embodying Analysis across the Qualitative Continuum

Doing Legwork: On Thinking through Data with Mind-Altering Medications

I swallowed another glug from one of my ever-present cans of Diet Coke, carefully setting the cold, sweating can on a coaster on the dark wood coffee table. Sighing, I turned my head and looked over the living room and entryway, seeking relief from the combination of boredom and agitation that always sets in after surgery. I was already sick of the view from our L-shaped couch where I sat for almost all of the 15 or 16 hours a day that I wasn't asleep. A veteran of 16 other operations during and in the years following treatment for bone cancer in my right leg, I was used to the twitchy feeling in my body when MS Contin, one of the few narcotic pain medications that didn't make me unbearably sick, fuzzed the edges of my world and simultaneously caused a weird restlessness in my legs.

Well, I thought to myself, I guess it's a weird restlessness in my *leg*. Singular. I looked down at the small remaining piece of my right thigh, ensconced in a cast and immediate post-operative prosthesis (or IPOP), which was basically a metal rod embedded in a cast at one end, with a primitive prosthetic foot at the other end. My prosthetist had informed me that the polite, medical term for my remaining bit of thigh is "residual limb," rather than stump, but I rejected both of those in favor of my husband Glenn's coining of the term "leglet."

Sighing with great, unwitnessed drama, I began to read again through the analytic memos I had written linking quotes from interview transcripts and participants' written narratives about communication among aunts, nieces, and nephews within extended family and chosen family (voluntary or fictive kin) relationships. I tried to concentrate on the ways in which aunt roles fit within the lives of women who (whether by choice or circumstance) did not have children of their own, in order to articulate a coherent and meaningful theme. Time and time again, I delved into the topic, only to be distracted by physical sensations—dizziness, pain at the surgical incision site or in my phantom limb, numbness and "pins and needles" in my bottom, already sore from too much time on the couch since returning home from the hospital two weeks ago.

"Damn it!" I muttered to myself as I lost my train of thought yet again. Hot tears pricked at the corners of my eye lids, and I closed my eyes briefly, consciously taking a deep breath to slow my rapid, nearly panting, respiration. I became aware of the tension headache that was building at my temples and the tightness in my neck and chest. The low throb at the end of my leglet continued through the dampening veil of MS Contin that kept the agonizing pain at bay. I tried and tried to ignore my troublesome body in order to get some work done.

What would happen if I actually paid attention to my body? I wondered.

Opening a blank document to do some free writing, I pondered my body and the bodies of aunts, nieces, and nephews. Fingers dancing across my laptop, I wrote about pain and discomfort, and then physical needs more broadly—children needed to be fed, cared for, played with, picked up and dropped off at school, sports, piano lessons, play dates. I thought of the bodies of aunts who had not borne babies yet had arms that hugged close the bodies of the nieces and nephews in their lives, and those aunts raising their own children who formed a net of love and security for their nearby nieces and nephews. I thought of aunts who lived far away from their nieces and nephews who maintained close ties with periodic visits, treats, and adventures, and of disliked aunts who caused bodies of nieces and nephews to stiffen, pull away, and frown.

I stopped typing, having lost my train of thought once more. Sipping my cold, bubbly Diet Coke, I noticed again the discomfort in my buttocks, the low throb of my severed femur, and the lessening of tension across of my forehead that evidently eased somewhat while I wrote. The cognitive fuzziness caused by the medications made it difficult to follow a sustained line of thought, causing me to cycle through the same data and memos over and over again. Yet the connections lurching between my muddled mind, my uncomfortable body, and my data analyses also prompted insights about the embodied nature of aunting relationships (for more on aunting relationships and related cultural discourses, see Ellingson & Sotirin, 2010; Sotirin & Ellingson, 2013).

Sipping more Diet Coke, I contemplated my body as a site of analysis. Since completing my M.A. thesis, I have not written a significant-length scholarly paper or presentation without being under the influence of prescription-strength pain medications. I live with "late effects" of cancer treatment, which are the physical and psychological/cognitive conditions that remain after completing cancer treatment, or which begin in the months, years, or decades following treatment as a result of chemotherapy, radiation, surgeries, medications, and other treatments (Lund, Schmiegelow, Rechnitzer, & Johansen, 2011). In my case, late effects of treatment were arthritis in my knee, fractures and infections in my femoral bone graft that necessitated further surgeries and other treatments, difficulty walking, and chronic pain. Following the eventual amputation of my leg above the knee (almost 20 years after my cancer diagnosis and treatment) during my recovery from which I took narcotic pain medication, phantom limb pain compelled me to begin a new regimen of neuropathic pain medications (e.g., those used to treat pain due to diabetic neuropathy and fibromyalgia). I also began exploring acupuncture, reike, massage, biofeedback, and other complementary treatments for phantom pain.

Pain medications are powerful actants that co-construct my way of being in the world, including with data and analysis processes. My body-self, pain medications, prosthetic leg, and many elements of my physical environment within my home, workplaces, and in public, intra-act in complex ways to produce my body-self, pain, cancer survivorship, the pharmaceutical industry, and more. My particular body-self really knows little about a hypothetical world in which I conduct data analyses without the influence of pain medications (and pain). I tend to think that my analyses without the veil of pain medications would be somehow better, but this is arguably unlikely. Despite some fuzziness of thought, I function well enough to work at a full-time faculty position, drive a car, buy groceries, and generally fulfill my responsibilities; in other words, I am not incapacitated by the medications, only inconvenienced, distracted, annoyed, slowed down, and with less endurance than I otherwise would enjoy. The increased number of iterative passes I make through my data because my concentration is limited and I am distracted by pain undoubtedly lead to *different* conceptualizations of my data than I would otherwise form (and that would form me). But the hypothetically non medicated conceptualizations would be not necessarily be any better or worse, just *other*.

Sighing one more time, I returned to my piles of memos and transcripts, and began once more to reflect on aunts, nieces, nephews, and bodies. Capitalizing on the caffeine boost from my soda intake, I focused on a whiff of an idea that hovered on the edge of my consciousness—about rituals that make aunties into chosen family—and I tried to render it in language.

In this chapter, I contemplate the ways in which researchers' and participants' body-selves are implicated in data analysis procedures. I discuss data analysis as a material practice, as always already embodied, and as analysis located in the head, heart, and gut.

Data Analysis as Material Practice

Qualitative researchers in the early stages of data analysis achieve "intimate familiarity" with their textual materials by rereading them many times, making notes on emergent trends before proceeding to later stages of analysis (Warren & Karner, 2009). While researchers may think of this process as getting the ideas into our heads, we engage in an embodied process of intimate familiarity through our bodies—we read data, listen to recordings, view photographs, maps, or other images, make notes with our hands, and so on. Our whole bodies process our data. Konopásek (2008) explained that "analytical work is in an important sense a material praxis (and vice versa)" (¶30). Likewise, MacLure (2013) suggested that scribbling in margins, shuffling papers, and underlining printed data constitutes an embodied process, a connection to the materiality of data and of the entire analysis process which is, after all, accomplished with hands, eyes, ears, shoulders, and back, the lap that holds the laptop computer, and so on.

Using computer programs that assist with qualitative data analysis, "we can *create, see* and *manipulate* various objects. These objects can be of different sizes and shapes; they can be hidden, moved, split, colourised, grouped, and regrouped, forgotten and rediscovered on unexpected occasions" (Konopásek, 2008, ¶20; original emphasis). Similarly, those who have used or continue to use printed paper copies, colored pens and pencils, scissors, paper clips, and so on for data coding and manipulation create new objects (i.e., groupings of quotes and notes) within a "textual laboratory—which has the power to shrink time and space distances between observable phenomena so that everything important is present and under control" (Konopásek, 2008, ¶22). The grouping, networking, coding, and commenting on of quotes enables researchers to visualize connections among ideas, deeply impacting our continuing construction of meaning.

The physical rearrangement of documents used in analysis—different types of data, analytic memos, and reflections, notes on research processes, even to-do lists, help researchers think through their analyses. Explained one researcher:

> I moved data around, generated queries around "codes," and re-arranged the piles to re-engage my memories of my field experiences. These material practices, pen to paper, hand moving to underline and write, "doing," were a necessary part of my analytic practice. . . . The promiscuous materiality of analysis came alive through this affective engagement that provided a way to (re)engage the bodily and affective conditions of research.
>
> *(Childers, 2014, p. 821)*

Rather than merely housekeeping chores or computer clicks, data analysis is grounded deeply in the material world. Researchers' choices about organizing and handling our data materials *matter*, and they should be carefully considered in terms of fit with researchers' personality and preferences, their participants' capacities and needs, and the types of data with which researchers are engaged.

Analysis as Always Already Embodied

Whether coding with an established set of codes, inductively deriving a typology of categories or narrative schemas, reading data through a theoretical lens, or reflecting on fieldnotes in order to construct an ethnographic story or poem, all along the art-science continuum, researchers still must make sense of their data or other empirical materials. And this sense making can benefit from conscious efforts to thoughtfully attend to researchers' and participants' embodiment. I offer several concepts that are helpful in understanding how bodies are implicated in data analysis.

First, postpositivist understandings of coding and structured qualitative analyses limit the results of qualitative analyses to patterns that are presented as fixed and finalized. Some researchers argue that coding is not a useful mode for qualitative analysis and should be abandoned by interpretive, critical, and poststructuralist researchers (e.g., Lather, 1991). For example, St. Pierre and Jackson (2014) argue that to code necessarily involves understanding written data—generally interview transcripts and fieldnotes—"as brute data waiting to be coded with other brute words . . . [C]oding . . . is thinkable and doable only in a Cartesian ontological realism that assumes data exist somewhere out in the real world to be found, collected, and coded" (p. 715). Coding also enacts researcher power over participants' words, put succinctly: "Researchers code; others get coded" (MacLure, 2013, p. 168). Others, particularly social constructionists (Charmaz, 2006) and feminists (Hesse-Biber, 2007), point to the significant value of discerning patterns in qualitative data (despite its limitations), particularly when researchers' standpoints are carefully considered as part of analysis.

Researchers can shift their thinking about codes, themes, narrative structures, or whatever patterns they seek in their data from conclusive, rigid, finalized categories or containers to a more flexible and open-ended sense of findings without abandoning qualitative analyses or ignoring the embodied aspects of analysis. Following Deleuze and Guattari (1994), the concept of "becoming-analyses" includes always new interpretations and meanings, viewed from different times and places and spaces (Holmes, 2014, p. 783). To some extent, this can be understood as a furthering of the constructionist approach to grounded theory advocated by Charmaz (2006), wherein the constructed nature of findings, the standpoint of the researcher, and the relationships among participants and researchers are taken seriously in the development of results or outcomes, which are acknowledged as partial and contingent. Building on the acknowledgment of the constructed nature of research reflects the middle ground of the art-science continuum perspective as it intersects with critical theory.

Moving a bit closer to the artistic end and engaging a tangential line of flight, Markham advises researchers to change their analytic agenda "from one that seeks stability or order to one that seeks to compel, relate, or explore, understanding the inherent open-endedness of this act in contextual space and time. . . . [and] acknowledging that one is engaging in sense–making rather than discovering" (Markham, 2013, p. 8). If findings are not only constructed (rather than discovered) but also always becoming, one vital element of that unfinished, contingent quality is embodiment. This shift in mindset from final to open makes room for the dynamic and ever-changing nature of bodies in the world—researchers' and participants'—and the structures that surround them. Bodies too are dynamic, and when we embrace embodied analysis as influenced by concepts of openness and becoming, researchers can mobilize our bodies to engage in any approach to data and sense making that fits with our goals and processes. That is, I contend that there is no need to choose between coding and becoming; data analysis assemblages can accommodate a wide range of processes, concepts, and body-selves, including holding space for philosophical contradictions without resolving them.

Second, attention to embodiment may be facilitated by practicing methodological unfaithfulness, both in terms of drawing from a multitude of approaches and in terms of not staying true to the stricter parameters of methods. Of course, such transgression must be carefully accounted for and its implications considered, but it nonetheless is rich in possibilities. Indeed, coding, "when practiced unfaithfully, without rigid purpose or fixed terminus, the slow work of coding allows something other, singular, quick and effable to irrupt into the space of analysis. Call it wonder" (MacLure, 2013, p. 164). Wonder is embodied, "simultaneously Out There in the world and inside the body . . . distributed across the boundary between person and world" (MacLure, 2013, p. 181).

Likewise, Childers (2014) boldly reclaimed a gendered, sexualized term—methodological *promiscuity*—rife with pleasure, erotic power, and the edginess of a woman stepping out of bounds.

> [M]y approach to analysis became promiscuous. Grounded theory, situational analysis, pleated texts, rhizomatics, policy analysis, and discourse analysis were suggestions and flexible tools rather than recipes. . . . I was doubly promiscuous, engaging in conventions that might be the very source of analytic containment, yet breaking that containment by (mis) appropriating them. The promiscuous materiality of analysis came alive through this affective engagement that provided a way to (re)engage the bodily and affective conditions of research.
>
> *(Childers, 2014, p. 821)*

Like Childers and MacLure, I embrace unfaithfulness and promiscuity in methodology, which I constructed (with overlaps and differences from their approaches) as crystallization (Ellingson, 2009a). Our shared emphasis on embodiment in analyses is echoed in a "blue-collar" voice by Saldaña's (2014) "rant" about his own belief in maintaining a large theoretical and methodological tool box.

> Me, I keep myself open to bein' and doin' what needs to be done. I'll be a grounded theorist when I need to be. I'll be a statistician and crunch some numbers when I need to. And I'll be a poet or playwright or artist when the occasion calls for it. It's all 'bout findin' the right tool for the right

job. . . . Blue-collar folks are good craftspeople—they know their tools. They work hard and they sweat harder. Sweatin' analysis is gittin' your hands dirty in the data.

(Saldaña, 2014, p. 978)

I appreciate Saldaña's invocation of embodiment through the sweat and dirt of analysis and the spoken dialect of his youth (see also the generative discussion of analytical pluralism; Clarke et al., 2015). Each of us in our own ways provides embodied methodological paths out of the modernist, positivist land of theoretical allegiances and methodological rule-following. Moreover, none of us require outright rejection of any paradigm, method, or practice as the cost of making the trip; researchers can each pack their own bag for the journey.

A third helpful concept for doing embodied analysis is to attend to the "strange relations" within data; "the strange relations are coding's uncanny 'other'" (MacLure, 2013, p. 180). Researchers can pay attention to all the things that are omitted from the coding process—"movement, difference, singularity, emergence and the entanglements of matter and language" (p. 171). The parts of our data sets that resist coding or categorization can be embraced rather than ignored or minimized, which MacLure calls "hot-spots," using the metaphor of temperature to name that which burns bright in our body-selves, urging us to pay attention to the strange relations we discern (p. 172). An example of attending to strange relations in data through embodied analysis is articulated by Roberts (2013) in her analysis of the performance of black and brown bodies in dance and their enactment of racialized history and discourse.

Turn the pathologizing cognitive discourse in my head on its head, shake it out and create a space for accumulated, stagnated history, experience, and affect to cascade and circulate in and through our bodies as a coperformance. This deeper, more demanding work will require me to engage my body/put my body in it together with my mind as an integrated site of interpretation.

(Roberts, 2013, p. 281)

Roberts documents her embodied response to the performance of dance she witnessed—a sense of weight, held breath, tension in parts of her body—as she reflects on her body and the body of the dancer entangled in strange relations brought to light during data analysis. Hot spots in data can spark attention to the researcher's body and the researcher's body can spark attention to a hot spot in the data; the hot spot and the researcher's body-self intra-act and open up space for developing significant insights from one's data.

Fourth, think of analyses as necessarily attending to the discursive and the material simultaneously by "think[ing] data differently" in order to reject the material/discourse dichotomy (Jackson & Mazzei, 2013). Researchers can join analysis of cultural discourse with analysis of individual experience. Drawing on Barad (2007), they argue that researchers rely on a false dichotomy between the social and the natural and suggest that interviews instead be thought of "as discursive, as material, as discursive *and* material, as material<->discursive, and as constituted *between* the discursive and the material" (Jackson & Mazzei, 2013, p. 269, original emphasis). The researchers offer a moving example of a black professor whose small university office became a safe space for black students to hang out and be mentored. At the same time, some white students felt excluded from her office. The issue was both discursive and material: students and professors all had bodies and agency, and all were acted upon; discourses of race, racist legacies, and white privilege in universities circulated in and through bodies; and the materiality of the office was consequential, demonstrating how the discursive and material forces intra-acted to produce a space that is experienced as a safe space by black students and a place of exclusion by white students. In this example, the inseparability of the material and the discursive become evident in and through the racialized bodies of students and the professor operating within a sticky web of oppression and privilege (see also the articulation of genealogical-phenomenological schematization as a way of bridging "analysis of systems of knowledge" and individual accounts of lived experience; Bryson & Stacey, 2013). Jackson and Mazzei's process of thinking data differently demonstrates that researchers can meaningfully challenge not only the discursive/material dichotomy but all dichotomies that force people and actants into one of only two categories, despite the ample evidence that the rigid distinctions between the categories cannot hold up in the face of the lived experience of an extraordinary variety of bodies.

Fifth, consider the notion of "plugging in" one text to another as embodied process. Drawing on Deleuze and Guattari's (1987) notion of "plugging in" to name a process of thinking data through theory and theory through data, plugging different texts into another which "creates a different relationship among texts: they *constitute* one another and in doing so create something new" (Jackson & Mazzei, 2013, p. 264). Plugging in, or considering theoretical concepts with data in an assemblage approach, involves three "maneuvers"— disrupt the theory/practice dichotomy; show how a theoretical concept, when "plugged in" to a text, generates (or makes possible) particular questions; and repeatedly go through the same "data chunks" to deconstruct and rethink them through the process of plugging in (Jackson & Mazzei, 2013, p. 264). The plugging in process is a machine metaphor, and this connection illuminates the process specifically as a material one. That is, the process of plugging of one text into another is accomplished through the body as much as it is the mind. A great example of how girls' bodies, subjected to sexual regulation by culture and their schools in particular, show up in data, such as this "hot spot" (MacLure, 2013) that Ringrose and Renold (2014) identify in their study of slut-shaming and girls' sexuality. Some of the girls had attended the "SlutWalks" event with their mothers and brought back the concepts and shared them with others who had not been able to participate in this event and thus

> were cut off somewhat from its potential and possibility for rupture. "Plugging into" (Jackson & Mazzei, 2012) the SlutWalk and bringing this energy back into the school was a rupture in time-space. But the complex stickiness, and simultaneous allure and revulsion of "slut" lives on in a powerful way for teen girls because of the way their bodies are coded and assembled.
>
> *(Ringrose & Renold, 2014, p. 777)*

Researchers can "enter the assemblage" of their data in order to explore the complex interactions of the materiality of the girls' bodies, the discourses of SlutWalk and of local and global "slut-shaming" of girls and women, and the ways in which "plugging into . . . the SlutWalk" became part of (some of) the girls' experience directly and others by association, read by the researchers through feminist theory that is also part of the data assemblage. Plugging in different theories, concepts, or texts into data do not just interpret the data but also constitute the theory, concepts, or other texts through intra-action. Plugging in a series of concepts is a great way to think through the entanglements of the material with the discursive in generative ways that open up space and create new connections.

Sixth, embodied data analysis can be thought of using the concept of abduction. Like most social researchers, I was socialized into an acceptance of quantitative methods as invoking deductive logic and qualitative methods as inductive reasoning. Yet the qualitative method is also typically described as iterative and cyclical, involving repeated considerations of theory and concepts along with ongoing inductive analyses of data, a process that can be considered more abductive than inductive. Abduction

> is, in every sense, a means of inferencing. It is precisely in this quality of being a 'means-of-inferencing' that we find the secret charm of abduction. On the one hand it is a *logical* inference (and thereby reasonable and scientific), and on the other hand it extends into the realm of profound insight (and therefore generates new knowledge).
>
> *(Reichertz, 2004, p. 159)*

In qualitative research when we use inductive analysis, our "aim is not to pretend that the researcher does not have pre-existing theoretical ideas or emotional responses related to the research topic. Rather, the aim is to engage theory and emotions reflexively" (Ezzy, 2010, p. 167). The concept of abduction may be energizing to some researchers as they do embodiment. Abduction connotes a sense of movement and play, like a pinball bouncing off a variety of surfaces, whirling about with kinetic energy, a metaphor that indicates how nonlinear, unexpected, and often pleasurable the process of making and remaking connections can be.

Researchers do not make these connections among data, concepts, theory, and published research results merely in our minds but feel them throughout our bodies as they form—my heart races, my spine straightens, my eyes sharpen when a connection is made; my shoulders slump, my lungs deflate, my mouth pouts when a tentative link skitters just out of reach. Researchers do abduction with our bodies, and rethinking induction as abduction may enliven data analysis processes.

A seventh useful idea is to explore embodied metaphors and imaginative language to describe our processes of data analysis. The way in which we explain how we conduct analyses influences how we understand and experience our actions and intra-action. Explains one qualitative researcher:

> Analysis is an off-road journey I take in slow, up close ways. . . . I have to care for, prepare, and sustain my mind–body for analysis . . . [by] reviewing data, text I am writing, writings of others, and sometimes art that resonates. Taking impressions of these with me, I will myself to work through/with/in them.
>
> *(Daza & Huckaby, 2014, p. 804)*

Do researchers find patterns that are scattered throughout data like shells in the sand? Do we build our patterns out of planks of data? Do we simmer our "raw" data into a well seasoned analysis? Or we can be archaeologists, discovering bits of data like artifacts in the field, hauling them home, sorting them, and cleaning them up to make their details more clearly visible? Devising an apt metaphor for one's data analysis processes may be generative.

Head, Heart, and Gut Analysis

Researchers do not merely use their bodies to access knowing in their minds; our knowing is interwoven throughout our bodies (see Chapter 1). Barnacle (2009) argued that our "gut feelings" are not merely metaphorical but crucial sensations intricately involved in sense making.

> Being critical, therefore, becomes an aspect of how one lives one's life, and this is not reducible to a specific skill set that can be deployed or withheld at will. A gut, engaged "moodfully" with the world, to borrow from Heidegger, offers a better model for describing such a phenomenon than a conception of mind dominated by a calculating brain.
>
> *(Barnacle, 2009, pp. 31–32)*

Knowledge grounded in bodily sensations encompasses uncertainty, ambiguity, and messiness in everyday life; it is inherently and unapologetically subjective, celebrating—rather than glossing over—the complexities of knowledge production. "Our bodies are analytical terrain, not simply vessels for the mind" (Daza & Huckaby, 2014, p. 801). Thus our gut feelings and other bodily sensations arise as we reread and analyze our data—tears, muscle tension, headaches, feeling energetic, smiling, trembling—and give us clues to meanings embedded in our data (itself a construction of our and our participants' bodies). Bodén (2015) described the sensation of being pulled toward particular pieces of data: "Something dragged me back to the situation, it sparkled and glowed . . . charmed me, and discomforted me" (p. 193). Such clues are complex and require careful reflection and interpretation. Reflecting on her analysis process, an ethnographer explained: "My body and flesh are out of joint with the sedated coding that I already anticipate[;] . . . thought relates to my flesh and not my brain" (Holmes, 2014, p. 784). Some data live in researchers' bodies, as though bits had

> been ingested into my blood stream and body's fibers. . . . Some curious fragments seep through my pores, in molecular ways becoming part of my flesh, (de)composing with my body, necessarily living with and in me, entering a new kind of fleshly decay and analysis that goes beyond coding.
>
> *(Holmes, 2014, p. 783)*

If we consider data analysis a manifestation of critical thinking, it follows that researchers must employ our guts as consciously as possible and consider those of our participants, as we seek to sort through, discern patterns, construct coherent categories, typologies, theoretical readings of 'hot spots" and otherwise (re)assemble data into new forms. We need to pay attention to gut knowing as we analyze data through embodied sense making. Moreover, during analysis of qualitative data, researchers' bodies become immersed in textual data and make connections, which "involv[es] the goal of pulling together the strands of its meaning . . . A felt sense of the strands is present in our bodies. When we direct our attention to the felt sense, it gives rise to memories, associations and images" (Rennie & Fergus, 2006, p. 494). We have gut and intuitive senses of what fits, what works, what captures the meaning we are seeking or that seeks us. At the same time, researchers best remember that the gut and hearts are certainly not infallible: "Sensibilities can orient thought in ways not necessarily beneficial and can also limit or truncate one's openness to inquiry" (Barnacle, 2009, p. 32). For example, our understanding of emotions and accompanying physical sensations are subject to cultural "feeling rules" into which we are socialized, which are highly gendered, and which we generally invoke without conscious reflection (Hochschild, 1983).

Rationality and emotionality are intertwined, not opposites or even separable. "Analysis does not end, but rather begins with the recognition of [researchers'] own emotion" (DeVault, 1990, p. 105). One such embodied emotion is empathy. Explained one researcher about her empathy with the teenage girls participating in a study of sexual experiences: "Empathy is a hidden part of my method, but a junior scholar sitting at a conference table with senior faculty does not want to be accused of reading her own sex life into the evidence"—even when that insight about female arousal and lubrication (and lack thereof) has crucial implications for what participants are and are not stating about condom usage (Horner, 2010, p. 243), and for the intervention that will be designed based on this research. She points out that researcher squeamishness about, in this case, intercourse but really any messy aspect of the embodied being, such as elimination or rashes, leads researchers to miss out on important aspects of what our data is doing. Of course, researchers must be reflexive about emotions in our bodies and participants' bodies as we analyze data to make sure that we are not subsuming participants' emotions under our own emotional connections (Woodby, Williams, Wittich, & Burgio, 2011, p. 831). On the other hand, at times during analysis, feeling participants' intense emotions again can be exhausting and distressing. Knowing data in our heads, hearts, and guts is an emotionally taxing experience, yet a rich one for researchers, particularly when research addresses sensitive topics such as death and dying.

Practices for Embodied Analysis

Move Around to Move Your Analysis

No discussion of doing embodiment in qualitative research would be complete without mentioning the (admittedly well known) information that physical movement helps to spark creativity, relieve stress, increase energy, and improve mood. So regular exercise and even purposeful, meditative movement during data analysis work may help your analysis processes. Put more eloquently by two researchers, "[m]ovement meditation works to un-trap analysis from today's imagination;" it is "a physical contradiction to the inertia of seat-work" that will contribute to your data analysis and reflexivity (Daza & Huckaby, 2014, pp. 804, 806). When you feel stuck or feel like your assemblage has disassembled itself, move your analysis through your body physically. I solve analytical paralysis by vigorously doing housework with my ear buds in, listening to digital audio books that stimulate my brain with engaging narrative. One colleague of mine dances around her house to a few of her favorite songs, finding the beat of the music and letting her body flow to its rhythms. A brisk walk around the block or across campus might work for you; let yourself experiment to find movements that move your body-self to analysis by moving your body-self through time and space.

Mutual Embodiment

Researchers must attend to "how the interviewer's own embodied subjectivity interacts with that of the respondent in the mutual construction of meanings/bodies" (Burns, 2003, p. 232) in order to consider the complexities of meaning making. Attending to embodiment requires attention to the data as encompassing meanings tied to specific interactions in which specific bodies encountered each other intersubjectively. For example, Razon and Ross (2012) are both American-Israeli women conducting research in Israel. As they interviewed participants, some of whom were Jewish Israelis, others Bedouin, or Palestinians (who are Israeli citizens), and Jewish people from the U.S. living in Israel, they spoke in Hebrew, Arabic, and English, depending upon the people they were interviewing. The question, "Where are you from?" kept being asked as participants sought to locate the researchers' identities, ideologies, and loyalties through ascertaining where they were born, where they now lived, and what groups they identified with. At the same time, the researchers were investigating issues related to Jewish–Arab dynamics, and interviewed participants about their own identities and locations, each of them foregrounding and backgrounding aspects of their bodies, their education, and their multiple identities during the conversations. This attention to the mutuality of embodiment in the interview has implications for doing analysis. The same attentiveness to aspects of both researchers' and participants' embodied identities practiced by Razon and Ross can be brought to bear during analysis. The goal is to frame encounters and meanings not on the basis of what the participant's identity was or what she meant (by a specific comment) but who the participant and researcher are as body-selves *together* and what meanings their body-selves generated *intersubjectively* during their interactions. Continually returning to mutual embodiment keeps the focus of analysis on the betweenness of knowing and ways in which bodies are porous.

Harness Materiality of Analysis

Whether or not you embrace coding, reading data through critical theory, autoethnographic storytelling, or any other type of analysis, you will need to manage and interact with your data. And for all qualitative researchers I know, the amount of data ranges from significant to massive. Because the volume of transcripts, fieldnotes, documents, journals, visual data, and memos is tremendous, some sort of approach to getting familiar with all of the data and putting your hands on a piece of data when you want it becomes necessary. Some researchers store all their data electronically, some print everything, others use a hybrid approach. Regardless of what "system" you prefer (including a "system" of messy stacks of papers and folders cascading across a desk), you can harness the material aspects of data sorting, rereading, and reflecting as part of your embodied analysis processes by engaging in the physical acts of moving, cutting, pasting, copying, highlighting/underlining, and making notes—all of which can be done in either in virtual space or in material time/space. So if you like office supplies—colored pens, sticky notes, folders, paper—stock up. If you like electronic data, then invest in qualitative analysis software and learn how to use it. Consider getting flip charts (like giant sticky notes) and hang them on the walls to surround yourself with visual representations of connections between themes, people, and actants, and lists of ideas, concepts, tasks, and so on. Or lay papers and notecards out all over the floor or a large conference table, moving them around as you make connections. If you have the luxury of a dedicated work space, do a little (re)decorating of the space to make it better reflect your personality. If you work wherever you can find space, get yourself a snazzy cover for your laptop and/or a great computer bag, tote, or messenger bag to hold your work and supplies.

Harness E-Materiality of Analysis

Change the background colors and highlighting colors on your computer screen. Save your original data files (and back them up) and make play copies in which you can cut, paste, highlight, and annotate to your heart's content by pointing, clicking, dragging, and typing. If you like being able to rearrange files and ideas as you are writing, consider Scrivener software (available for purchase and download online). Qualitative analysis software includes ATLAS.ti, HyperResearch, NUDIST, NVivo, and MAXQDA. Most of these software packages have capacities for linking files that enable "embodied multimodal coding and analysis," that is, software provides an "ability to revisit a moment in a multimodal way" by hyperlinking documents, photos, videos, drawings, maps, and other texts to create connections among elements of multiple texts, highlighting the material and embodied aspects of experience by pointing to resonances across images, participant language (transcript), description (fieldnotes), and so on (Ntelioglou, 2015, pp. 91–92). The visual juxtaposition of images, sound, and written text may spark intra-actions that would not have happened without the collapsing of time and space accomplished by putting disparate moments together (Konopásek, 2008). The more you can make the work and play of analysis physical, the more opportunity you have to engage your body-self with your body of data and imaginatively foster intra-actions.

Bodies in Data

While some aspects of embodiment may seem readily apparent in your data, you may also find it helpful to read data with an eye toward detecting more subtle cues of bodies intra-acting with other bodies, actants, places, and discourses. I found a couple of helpful approaches to attending to subtle embodiment in analysis. First, Perry and Medina (2015) offer four excellent questions as a framework for analysis that accesses or constructs a variety of aspects of embodiment:

"What is represented through texts [data] (spatial, physical, verbal, etc.)?"

"How are cultural norms, histories, and knowledges inscribed or disrupted in relation to emerging narratives and identity constructs (such as gender, race, class, etc.)?"

"What relationships and dynamics (affects and forces) can be observed between bodies, between positions, material and immaterial contexts, instruction and action?"

"What is happening in this process (as seen through all of the above-mentioned foci)? What and how are changes, events, creations occurring?"

(pp. 6–7)

Keep these questions open in another window or printed out and placed by your computer or your folders of paper data, and return to them as touch stones as you reread, sort, reflect, and analyze. Another approach is to focus on how your participants (and yourself) engage in "somatic work" (Waskul & Vannini, 2008).

[S]*omatic work* refers to a diverse range of reflexive symbolic, iconic, and indexical sense-making experiences and practical activities. Through such experiences and activities individuals produce, extinguish, manage, reproduce, negotiate, interrupt, and/or communicate somatic sensations in order to make them congruent with personal, interpersonal, and/or cultural notions of moral, aesthetic, or [sic] and/or logical desirability.

(Waskul & Vannini, 2008, p. 54)

This concept helps researchers glimpse how the sticky web of culture is interwoven throughout our sensory experiences, socializing us to engage in somatic work to self-regulate and respond to others' sensory experiences. Riach and Warren (2015) borrowed the concept of somatic work to explore people's experiences of (their own and others') odors, as well as their meanings and attempts to manage them socially and symbolically within an office workplace; they conclude that smell is part of everyday (learned) social rituals for participants. Similar analysis can focus on the cultural work of sight, sound, movement, taste, and so on, using the somatic work concept to explore subtle aspects of embodiment.

Body as Analytical Instrument

We do analysis through our bodies. Investigate your body-self as an instrument not only of data collection but of data analysis; such reflections can help through "'auto-body-graphy,' that is mapping one's own em-bodied practices by asking: How is my body and imaginary, in particular, and how have bodies and imaginaries, in general, been shaped in my lived experience . . .?" (Daza & Huckaby, 2014, p. 807). The opening narrative of this chapter was an autoethnographic reflection on my reliance on pain medications, and their role as actants in my life, including the embodied way in which I engage in (and am engaged by) data analysis. At other points in my career, I have reflected publicly on my embodied empathy for patients undergoing cancer treatment (Ellingson, 1998), my aching leg in a crowded clinic with an insufficient number of chairs (Ellingson, 2005), my experience of the U.S. health care system as a long-term cancer survivor and amputee (Ellingson, 2009b), and my researcher body studying professional practices as embodied communication (and engaging in professional practice as an ethnographer) in a dialysis clinic (Ellingson, 2015), all of which were mutually constitutive with ongoing data analysis processes. At this point in time, I also can compare and contrast the representations to see how I understood my body at different times and under a variety of circumstances. I urge you to create an "auto-body-graphy" or other reflection on how your body-self has shaped and been shaped at different points in your life to have as a resource for understanding how your body functions as a research instrument.

Intersectional Analysis

Privilege, oppressions, and intersectionality of identities mean that "different bodies produce different knowledge differently" (Daza & Huckaby, 2014, p. 802). The points of intersection of identities are generative but also complex and may cause uncertainty when researchers try to determine how to engage intersections in analysis. As with all else in embodied analysis, there are no hard and fast rules, but I do have some suggestions.

- Don't implicitly or explicitly make an unmarked norm of white, middle-class, cisgender, hetero, able-bodied against which all other differences are constituted. Instead, frame all difference as inherently relational.
- Do look for how categories intra-act and are mutually constitutive, rather than separate, parallel forces (i.e., avoid the "additive" approach).
- Don't essentialize a group in the name of giving its members voice; all groups have within-group differences and are not homogeneous.
- Do resist seeing any category as dichotomous—part of intersectional analysis is to destabilize dichotomies and hierarchies of identity (Hoel, 2013).
- Don't feel that you must interrogate every intersection of every category; center your analysis on the differences that make a difference within the focus of your study.
- Do move back and forth between individual experiences, group or organizational dynamics, and structural discourses; keep them in productive tension to understand how the sticky web of culture is woven at all levels of social interaction.
- Don't feel that you must achieve a single, authoritative rendering of your participants. Instead, engage in sense making that emphasizes difference, is dialogic rather than synthesizing, and reflects mutual respect but does not harmonize.

Now that you have read my list, undo my do/don't dichotomies to read in between them. Feel the slipperiness of these guidelines; see how the absolutes cannot hold. Use them, think about them, and perhaps ignore them in order to participate in ongoing data analysis that takes intersectionality seriously (as an ethical imperative as well as a conceptual and analytical imperative) while not being overly constrained by it.

Map Bodies in Conceptual Space

Constructing maps of people, actants, and places can also help to illuminate ways in which bodies are—or should be—implicated in your data and analyses. Situational analysis features three types of mapping to offer nonlinear, spatial analyses as "analytic exercises" to complete during data collection and analysis (Clarke, 2003).

1. *situational maps* that lay out the major human, nonhuman, discursive, and other elements in the research situation of concern and provoke analyses of relations among them (see Figure 7.1);
2. *social worlds/arenas maps* that lay out the collective actors, key nonhuman elements, and the arena(s) of commitment within which they are engaged in ongoing negotiations, or mesolevel interpretations of the situation; and
3. *positional maps* that lay out the major positions taken, and *not* taken, in the data *vis-a-vis* particular discursive axes of variation and difference, concern, and controversy surrounding complicates issues in the situation.

(Clarke, 2003, p. 554, emphasis in original; see also Clarke's book-length discussion of mapping, 2005)

Clarke's work is well grounded in postmodern theory and offers a number of good exemplars. I have found the first type (situational maps) to be the easiest to generate and quite useful as I endeavor to keep bodies at the center of analysis.

As a starting point, a simpler version of mapping can be undertaken by taking a blank sheet of paper and writing key terms and issues in your data all over the paper (i.e., *not* in columns or rows), clustering those things that seem to cohere (even loosely) and then drawing lines to indicate intersections of categories, issues, structural and institutional factors, identities, etc. The idea is to use visual space to play with data and emerging ideas in order to (re)see connections and opportunities for imaginative inquiries into the constitution of lived bodies in places using a nonlinear format.

Reflexive Writing Journal

Writing on and through the body flows through researchers' writing most readily in open spaces that encourage creativity—journals and free writing on paper or computer work better than fieldnotes (over which the seriousness of the category

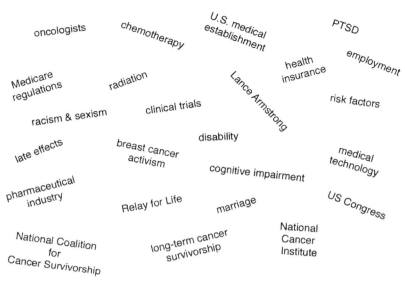

FIGURE 7.1 Situational Map.
Example of a situational map constructed from data on long-term cancer survivorship.

of "data" looms) or grounded theory memos (which constitute part of the formal analysis process). Keep a journal or file for writing that does not "count" as anything other than a place for reflexivity (which always sounds like a good thing to have accomplished). I urge you to consider making your journal handwritten, as there is some evidence that our brains process language and information differently when we use handwriting than when we type (see Cameron, 2002), so having more than one writing mode may be beneficial to reflexivity (although I say this hypocritically as someone with poor handwriting who types everything, but the point is still a good one). Drawing on Probyn (2011), Harris (2015) suggested that "[r]eflexive writing can evoke emotion, affect, and bodily attunement: the messy intricacies of the inter- and intra-subjective research process" (p. 1690). Continually reflect on emotion and your body in your regular fieldnotes, but also engage in regular, conscious reflection of how your body feels in the places you conduct research, how emotion is woven into bodily practices, and so on. Learning to switch back and forth between deeply attending to sensory experience and critical reflection takes practice. Researchers

> cannot feel fully and think about feeling at the same time. The challenge of "getting out of the head and into the body" . . . was limited by the need to constantly return to sociological thought and analysis. Reflexivity emerges as an embodied skill in the context of learned professional sociological skills.
> *(Gale, 2010, p. 215)*

Practicing reflexivity through journal writing should improve researchers' ability to attend to embodiment in themselves and others as they gain experience in articulating embodiment to themselves through writing that remains private and uncensored.

Transform Member Checking

Member checks or member validation asks questions about participants' understandings (Lindlof & Taylor, 2010). Member checks often reflect a more traditional, positivist-leaning stance of member checks as transactional, focused on assessing accurate presentation of the participants' subjective meanings. Yet participants' worlds are not static, and the experience of being in the study may inform participants' (evolving, embodied) understandings. "If truth is partially created during the interview and analysis processes, then it should not be assumed that an unchanging copy of this truth should be lodged in the participant's subjectivity" but instead is co-constructed in interview, and then again within member checking interview or other process (Koelsch, 2013, p. 171). In this way, knowledge is an ongoing (re)construction between researcher and participant. Therefore, it is more helpful to think of member checks less as a validation of what happened and what the participants think and more as an additional opportunity to talk with participants and collect their responses to researchers' accounts of their findings (Sandelowski, 2002).

Embracing creative forms of member checking can move the process from a dutiful task intended primarily for fact/interpretation checking (in a postpositivist mode), or a sincere but routine feminist-participatory ethic of respecting participants' voices in research results, to one that invites creative expression

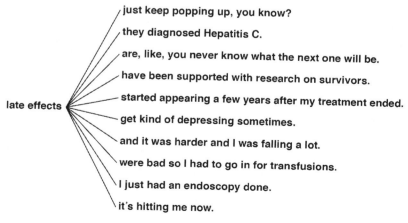

FIGURE 7.2 Word Tree.
Example of a word tree constructed from data on long-term cancer survivorship.

to complement existing data with imaginative, embodied representations. The additional perspectives collected can be in response to creative forms of sharing findings with participants that highlight participants' voices. For example, researchers in one project used interview transcripts to create "I-poems" and "word trees" made up of participants' words, and then had participants read them, underline lines or words with which they disagreed, and then discuss them with the researcher (Simpson & Quigley, 2016).

Such nontraditional summaries or points of entry to findings may help participants articulate their embodied experiences. Moreover, the back-and-forth of any member-checking interview constitutes an embodied session of meaning-making with further opportunities to try to elicit participants' heart and gut knowing, as well as more recognizably cognitive understandings.

Another researcher provided a transcript of their interview to each participant and invited the participant to make a "found poem"

> by going through the transcript, highlighting those sentences, phrases, or words that were particularly meaningful, powerful, moving, or interesting to them . . . They then took the highlighted sections and . . . reconstructed them into a poem that represented the thoughts and feelings they noted in their own interviews.
>
> *(Reilly, 2013, pp. 5–6)*

In addition to Reilly's found poem approach, researchers could adapt the "six word memoir" concept (Smith & Fershleiser, 2008) by asking participants to construct a six word summary of their perspectives or invite participants to draw, sketch, or collage an image that reflects their response to the research question(s) and to preliminary findings. Another idea would be to focus on a kinesthetic mode of knowing to ask participants to move their bodies in a dance, gestures, or other motions that express their thoughts and emotions about the ideas presented to them by researchers.

8

SPEAKING OF/FOR BODIES

Embodying Representation

Doing Legwork: Where Social Science Meets Art

It is to the artistic to which we must turn, not as a rejection of the scientific, but because with both we can achieve binocular vision. Looking through one eye never did provide much depth of field.

(Elliott W. Eisner)

I could smell brie cheese and something spicy as I stepped into the small reading room that had been reconfigured into a reception for the opening of our collaborative installation in the adjoining gallery space. A row of shiny metal chafing dishes gleamed in the harsh fluorescent light, their contents still hidden under dome lids. I was pleased to note an array of chocolate treats along a table at a right angle to chafing dishes. I nodded to my mother and to my husband Glenn, who sat in the front row of the seats set up for our opening panel discussion. My heart sped up as I watched people mill about the room.

Renee, my collaborator, arrived and set her bag on a chair, removing a small notebook. "Hi! Good to see you," she said.

"Hi! Everything is set up and ready to go," I said with a smile. "Kris is here, too," I added, referring to my former research assistant who had worked on the project with me and now attended medical school.

"Great," she said, smiling. Renee and I watched people move through the installation. I felt almost giddy. "This really came together, didn't it?" murmured Renee.

"It really did," I replied, looking at the poster with the artist's and researchers' statements, the screens with textile artwork on them, the computer set up to view

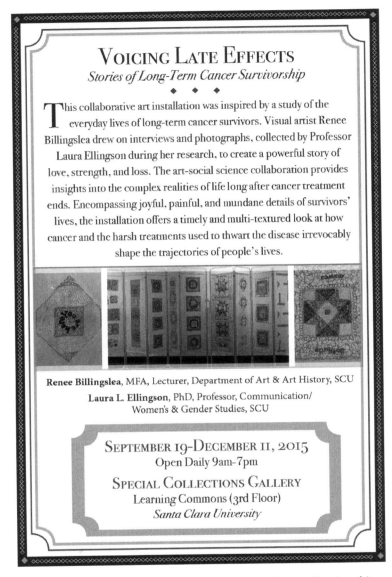

FIGURE 8.1   Voicing Late Effects: Stories of Long-Term Cancer Survivorship. This collaborative art installation used participatory photovoice and interviews to shed light on everyday experiences of long-term cancer survivors (Billingslea & Ellingson, 2015).

the online site that accompanied the installation, the paper art in a scrapbook on a table in a simulated waiting room, which people were flipping through to look at participants' photos and descriptions, and over at the informative displays about the construction of the textile art and about health problems associated with long-term cancer survivorship.

The panel began and I welcomed the audience. Then I read from our prepared statement, "About 14 million people in the United States are surviving or have survived the grueling treatments and triumphed over cancer. Yet life after cancer tends to be more difficult than might be expected. Most long-term survivors of cancer cope with *late effects* of cancer treatment, that is, the illnesses and chronic conditions that are caused by the surgery, chemotherapy, radiation, and medications used to treat cancer, as well as by the disease itself, that linger long after cancer is cured. My qualitative research explored survivors' experiences through the photovoice process [Wang, 1999], in which participants were invited to take photos of their daily lives and then to share their thoughts about the significance of each photo in an interview. I conducted traditional, social scientific analyses for scholarly publications [Borofka, Boren, & Ellingson, 2015; Ellingson, 2017]. 'Voicing Late Effects' is the outcome of a collaboration between myself and visual artist Renee Billingslea."

Renee smiled and explained, "I drew on interview recordings, transcripts, and the participants' photos to create the art installation, along with some wonderful student assistants. I dialogued with Laura about the research as our project evolved. I wanted it to be reminiscent of a medical environment, which is why we incorporated portable dividers used in mobile clinics and health fairs as the framework for the textile art. Then I designed panels for each segment of the dividers. Each one reflects ideas from the participants. I also worked with a student who helped create scrapbook pages using threads and sewn elements to resonate with the main piece of artwork."

I turned to Kris Borofka and nodded. Kris, a recent alumna and my former research assistant, explained cheerfully, "I helped Laura with lots of aspects of the research project, but my main focus was on creating a website that highlights survivors' stories and photos while also providing information about late effects and links to survivorship resources. It's available at through Laura's SCU faculty page."

Nodding, I picked up the threads of our presentation. "Blurring the boundaries between art and social science fosters deepened and complex understandings of social phenomena as researchers and artists combine the verbal and the visual, stories and analyses, individual experiences and societal structures. We urge boundary blurring work in which the social science not only inspires or is expressed in artwork, but that the production of the artwork is intricately interwoven with the development and refining of research analyses and reports. Creative representation of research findings invokes multiple senses in material formats to go beyond writing to include images, textures, sounds, or other sensory cues and attend to details of their own and participants' bodies in ways that bring them alive for readers and highlight their embodied being." For the next hour we took questions and socialized with the attendees, hearing people's cancer stories and reactions to the artwork.

What's In and What Lurks Outside

> There is no point in carrying out a carnal [social science] backed by practical initiation if what it reveals about the sensorimotor magnetism of the universe in question ends up disappearing later in the writing.
>
> *(Wacquant, 2009, p. 123)*

Qualitative researchers leave so much out of our tidy research reports, especially methods sections, a choice that obscures the ways in which their composition reflects and affects our bodies and those of participants (Ellingson, 2006). Omitted from our reckoning is all mention of "the *invisible work* that helps to make a research report. . . . the *uninteresting*, everything that seems to be not worth telling . . . the *obvious*, things that everyone is taken to know. . . . everything that is for one reason or another being *repressed*" (Law, 2007, p. 602, original emphases). Of course, much of what is deemed invisible, uninteresting, obvious, or repressed concerns messy, leaky, unpredictable, imperfect, and wonderful bodies. Even those research accounts that include narrative, poetic, or other arts-based representations intended to complexify conventional academic accounts, and those that use critical theory to unpack moments in our data, often marginalize the role of the body or simplify complex bodies into a singular category.

Transgressing normative writing conventions reflects bodies out of bounds and participates in the further co-creation of transgressive bodies. "To speak 'in the wrong way' not only reveals the speaker to be connected to his/her body, particularity, and context, but also risks blurting out that the emperor is naked—the emperor speaks from a body" (St. Pierre, 2015, p. 337). Yet the dominant ethos of what I refer to as "immediately-post-positivism," that is, acceptance of the impossibility of complete objectivity coupled with the retention of (most) positivist rules and criteria for research, reflects a tolerance limited to qualitative research that plays by the same rules and produces the same bodies. Such an approach to qualitative methods assumes that representation is an unproblematic incarnation of a reality. Within this paradigm, "[g]ood method creates a reliable representational conduit from reality to depiction. . . . But this is a slight of hand. This is because *realities are being done along with representations of realities*. (Law, 2007, p. 601, original emphasis). Thus "research does not 'represent' reality but rather indexes the various ways reality might be produced and how different ways of producing reality have different social, economic, and political effects" (Martin & Kamberelis, 2013, pp. 672–673).

Not only methods sections but all other sections of conventional accounts—results and discussion sections especially—leave out more than they can include. Of course, results can never directly contain more than a tiny fraction of the wealth of data gathered in qualitative studies. However, the elimination of the body from published accounts—pun intended—is a more insidious present-absence; that is, conventional accounts do not just leave out a lot of stuff; they exclude virtually all body stuff—the mess, the odors, the bodily failures, the embarrassing moments, the tears, the joyful laughter—and the ways in which our knowing is corporeally contaminated, rather than cognitively pristine. Conventional social scientific reports certainly are not the only format that risks marginalizing materiality, however. Critical theorizing may leave bodies so abstracted that they appear reduced to so much discourse—albeit often fascinating, nuanced discourses that reveal the intricate workings of power—almost incidental to the analyses. Limitations also arise in the ways in which autoethnographic and performative texts frame the body and its emotions within dominant discourses of oppression.

In this chapter, I offer several provocative perspectives on the positioning of researchers' and participants' bodies in research accounts across the qualitative continuum. Following that, I describe specific practices for doing embodiment in representation. However, before I turn to those topics, I want to briefly address my choice to retain the framing of this chapter in terms of issues of representation. Those qualitative researchers who are immersed deeply in one or more areas of contemporary critical theorizing may have noted a tension arising from my continual use of the term "representation" to describe the products of research, in concert with a heavy reliance on concepts indebted to poststructuralist and other critical perspectives that assume that representation of an external reality is not actually possible. Currently, "non-representational theory" has articulated the impossibility of producing representations of research that can capture or reflect "our self-evidently more-than-human, more-than-textual, multisensual worlds" (Lorimer, 2005, p. 83).[1] I continue to use representation largely because it is accepted terminology among qualitative researchers across the continuum whose work does not rely directly on critical theorizing, widely used in qualitative methods books and journal articles, and retained (even if problematized) by many researchers who use critical theory in their research accounts. Representation has a lot of baggage, but it is also useful, and I could not construct a suitable alternative that did not, in turn, invoke other significant constraints on my ability to speak across paradigms and bodies of theory.

Next I invite consideration of several perspectives on embodiment and representation, including: composition theory, queered and cripped; a critical geography (re)take on maps; subjugated knowledges and knowledge of subjugation; radical specificity of embodied experience; and refracting bodies through crystallization.

(De)Composing Bodies

McRuer (2004) brought together composition theory, queer theory, and crip (disability) theory to argue persuasively that "composing bodies" is both an embodied process—we write as body-selves, our perspectives are rooted in our bodies—and a result of the writing process—when we write, we join with other discourses, people, and objects to mutually construct the bodies of ourselves and others. Yet standardized expectations circumscribe the types of bodies that produce and are produced through normative writing practices. All bodies are expected to conform to naturalized ideals of "normal" bodies that are "continually produced from the disorderly array of possible human desires and embodiments" (McRuer, 2004, p. 51; see also Thomson, 1997). Academia and its enforcers focus on producing "order and efficiency where there was none and, ultimately, on forgetting the messy composing process and the composing bodies that experience it" (McRuer, 2004, p. 53).

Bodies that do not fit in dominant paradigms—queer, disabled, outside gender binaries, mixed race—produce transgressive composing practices and produce bodies-out-of-bounds. Although McRuer's argument is specific to university introductory composition courses, his linking of the writing process and its outcomes as manifested in and through the body is equally applicable to standard social science writing conventions. Such expectations are neither natural nor neutral, but reflect certain types of embodied experiences and in turn help to create researcher bodies that obscure difference and disorder (Coffey, 1999) in favor of "deceptively tidy accounts" of "what happened" during research projects, summed up in bare bones methods sections and orderly results sections of journal articles (Ellingson, 2006, p. 299). While researchers may be accustomed to thinking reflexively about their bodies, the link between the ways we write and the continual (re)formation of our bodies through our writing practices constitutes a further concern. As we produce and are produced, we have the opportunity to subvert dominant expectations in both subtle and explicit ways. Researchers can reimagine how we *do* representation in ways that uphold not only research norms but larger social discourses and structures, interrogating our "corpo-reality" and embracing "disruptive, inappropriate, composing bodies—bodies that invoke the future horizon beyond straight composition" (McRuer, 2004, p. 57). What would that future horizon look like? Resisting idealized bodies and supposedly universal research processes by invoking the workings and practices of individual researcher bodies opens up cracks in the hegemony of methods sections' key phrases to allow room for embodied resistance. "Cutting against the tired logocentrism of the universal speaker, impure voices powerfully remind us of the embodied mediation of 'the human': risky, always partial, ever somewhat wobbly, and necessarily contingent" (St. Pierre, 2015, p. 343). In a powerful example of cripped writing, Birk (2013) described writing from a body with chronic pain and a disabling condition.

> [I]t is also exceptionally difficult to write without a body that feels like one's own, to write from an internal space that is constantly and unpredictably assaulted by the chaotic circuitry of a body in trouble. The shots of pain here and searing aches there cannot help but to distract the writer's train of thought and so to punctuate the text in question. . . . All ideas arise from within the walls of the body. All thoughts are shaped by the contours

of our ultimate material condition. No idea or experience is free from the constraints of the absolute structures of skin, muscle, and bone.

(Birk, 2013, p. 396)

Birk writes in and through the broken, wounded body she has, not the impossibly ideal body, and her writing reflects her embodied way of being. So too do all qualitative researchers write from bodies that cannot match the idealized body. When we compose representations of data—whether conventional social science or narrative/poetic/performative (or combinations of these)—we are positioned at the intersections of privileges and oppressions, identities and experiences, always already grounded in the imperfect wonder of lived body-selves (see Boylorn, 2011; Myers, 2008).

The Territory is not the Map

Alfred Korzybski (1933) famously coined the expression, "the map is not the territory," a cliché that nonetheless bears further reflection as we contemplate representation of research findings. Representations can be thought of as maps produced by researchers so that others may understand where researchers have gone and what they experienced. But we must be careful that in our haste to separate the lived experience of the territory from the map that we do not unwittingly naturalize the map as an objective or neutral account. Instead we should wrestle with the knowledge that "an interplay of social and technical rules is a universal feature of cartographic knowledge. In maps it produces the 'order' of its features and the 'hierarchies of its practices'" (Harley, 1989, p. 6). Map makers put their own location at the center of the map, thereby legitimating and promoting their own perspective on the world(s) they map (Harley, 1989). Strategic choices are made in map construction; all abstract or theoretical representations of any territory help us to grasp a territory in ways that are always already partial, partisan, and problematic, and yet useful as well (Eisenberg & Goodall, 2004). For example, one study attending to the "soundscapes" of urban neighborhoods posited that

> [m]aps provide a starting point because they absolutely do not attend to sound and motion; they silence the city and check its movement, and in so doing make sense. As such, maps are an undoubted aid to understanding; they help us to know and navigate the city, to find our way through and to places. Yet, and by the same token, they are partial documents.
>
> *(Hall et al., 2008, p. 1027)*

The lack of sound and motion point to the notion that what is absent from maps is fundamental to one's experience of cities as generally noisy, bustling places; the partiality renders intelligibility and navigation possibilities, but it denies essential aspects of city experience in order to do so.

The lack of sound and motion in a map is not just distorting in an abstract or universal sense; the absences and choices of mapping privilege some bodies some over others. The methods of mapping are those of the master (Lorde, 1984), and they produce maps that have face validity in a normative world. That is, the maps appear legitimate, authoritative, and objectively real. "Much of the power of the map . . . is that it operates behind a mask of a seemingly neutral science. It hides and denies its social dimensions at the same time as it legitimates" (Harley, 1989, p. 7). For example, world maps and globes almost universally display the north on top and south on the bottom, even though this is not inherently more correct than displaying the south on the top, and certainly not necessary. Research indicates that standard maps perpetuate "north-south bias," with the "up" position being associated with good, power, and affluence (Brunyé, Gagnon, Waller, Hodgson, Tower-Richardi, & Taylor, 2012; Meier & Robinson, 2004; Nelson & Simmonds, 2009). The cultural centrality of this bias was illustrated in popular culture in an episode (Redford & Sorkin, 2001) featuring "Cartographers for Social Equality" on the long-running NBC network television series about the U.S. president and his staff, *The West Wing*, in which cartographers attempted to persuade the White House press secretary that an inverted map featuring the Southern Hemisphere on the top should be mandated in U.S. schools. In 2015–2016, the provocative set of a political comedy show, *The Nightly Show* on the Comedy Central network hosted by African-American comedian Larry Wilmore, featured a south-side-up world map as the backdrop.

Maps also produce bodies, in that they contribute to cultural norms, practices, and representations that form the sticky web in which our bodies are constituted. Bodies can be understood as "maps of the relation between power and identity" (Rose, 1999, p. 361). Qualitative maps are usually maps in a figurative sense; a poem or story or painting offers insights, explicitly or implicitly addressing bodies, and a high quality qualitative analysis adheres to inductive (or abductive) processes to interrogate embodied experiences. All of these map (parts of) an embodied reality that cannot be a single, coherent reality. Thus any representation we create of bodies—regardless of genre, whether implicit or explicit—is still a map of the topic, the people, and the places, including the researcher(s). This does not mean that researchers should not create maps. Rather, I advocate that qualitative researchers carefully construct useful, generative maps; that we acknowledge our partial, partisan, and problematic maps were produced by our bodies and co-produce our bodies and the bodies of others; that we do not confuse the maps with the territories themselves (or the territories with the maps), and that we do not forget that all territories (geographic and cultural) are continually in flux within sticky webs of culture.

Generally, the admonishment to remember that the map is not the territory is intended to resist reifying the map as unitary, fixed, coherent, or encompassing of all aspects of the territory. Yet it is not just the map's inherent failure to be equivalent to the territory it represents that matters. Indeed, researchers can also strive also to remember that *the territory is not the map.* That is, no authoritative understanding of the territory exists to elude capture by imperfect maps. Just as strategic choices are made in map construction, so too are all understandings of any territory always partial, always in flux, forming and reforming, including in how the bodies within territories are marked ethnically and geopolitically. All maps (i.e., any representation of our research data, analyses, and results) implicitly and explicitly implicate particular embodied states for those for whom readers' understanding is circumscribed by our (literal and figurative) use of lines, colors, and symbols (Harley, 1989). While I reject an extreme postmodern position of relativity, where all interpretations are equally valid in the absence of a single, authoritative Truth, I nonetheless advocate that researchers both embrace maps and view them with continual skepticism. Some maps are more credible, veri-fiable, generative, pragmatic, or otherwise valuable in their embodiment and emplacement than other maps, even though different criteria for quality must be applied to different types of research across the continuum of methodologies (Ellingson, 2009a; Tracy, 2010).

Drawing on the philosophy of Deleuze and Guattari (1987), some method-ologists have sought to rescue mapping from its positivist roots by delineating it from tracing:

> A tracing is a copy and operates according to "genetic" principles of repro-duction based on an *a priori* deep structure and a faith in the discovery and representation of that structure. Tracings are based on phenomenologi-cal experience that is assumed to be essential, stable, and universal. . . . In contrast, "mapping" charts open systems that are contingent, unpredict-able, and productive.
>
> *(Martin & Kamberelis, 2013, p. 670; original emphasis)*

This more fluid approach to mapping positions traditional maps and mapping processes as tracing and sets tracing aside as (failed) copies of territories. In this approach, reality is composed of "lines of articulation" that denote status quo arrangements of power and their construction of "normal" life, and "lines of flight," which represent opportunities to break with the status quo, suggest-ing ways of practicing resistance. This reimagined form of mapping, then, would resist tracing only lines of articulation (i.e., status quo) and would instead include lines of flight (i.e., openness, possibilities) in order to render "visible the multiplicity and creative potentials inherent in any organization of reality: the offshoots, the expanding root systems, the ruptures, and the detours that are continually producing new relations of power and all manner of *becoming(s)*" (Martin & Kamberelis, 2013, p. 671). Research may engage in meaningful praxis by participating in the imagining of new lines of flight, that is, new possibili-ties for social justice, egalitarian relations, dignity, and the meeting of bodily needs. Such a conceptualization opens up possibilities for maps that reflect more dynamic being and less false stability. Researchers can redeem maps from disembodied, positivist discourses by producing maps (research reports, narrative, artwork) that show possibilities of what could become, not purported copies of realities. For example, researchers could map an organizational communication structure, an extended family, or relationships in an institution, not just geographic spaces (see de Freitas, 2012).

Subjugated Knowledges/Knowledge of Subjugation

Autoethnography and other narrative and arts-based research approaches (Faulkner, 2016) often appeal to scholars whose experiences remain marginalized within the academy—e.g., people who are disabled, people of color, LGBTQIA community members, those raised working-class or in poverty, and women in STEM fields (Ellis & Bochner, 2000). Such forms lend themselves to expressing individual experience evocatively and to invoking cultural critiques. Autoethnographic exploration of lived experience is often held up as a potentially emancipatory method, in that researchers are writing about their own experiences of oppression, articulating their own standpoints rather than having researchers speak for and about them. Such approaches are not necessarily any more liberating than conventional forms of representations, however; "in representing themselves, those writing about themselves could replicate the very structures they seek to destroy" (Richards, 2008, p. 1724). Drawing on Foucault, Gingrich-Philbrook (2005) suggests that autoethnography frequently intends to focus on "subjugated knowledge—ways of knowing, lost arts, and records of encounters with power" (p. 298). That is, researchers representing their own embodied marginalization seek to reveal and reflect upon the knowledge inscribed on their bodies by dominant culture(s). Yet such normative narratives may "only provided knowledge of subjugation, serving almost as an advertisement for power," when what is needed are "the kinds of subjugated knowledges we don't get to see on the after school specials and movies of the week" (Gingrich-Philbrook, 2005, p. 312). Creative survival at the margins of society spawns specific skills, insights, and ways of being that when shared help audiences to see outside the taken-for-grantedness of heteronormativity, gender binaries, white privilege, and so on. But because we come to know narrative structures in dominant discourses, we come to recognize our own bodies through the logics of such dominant narratives, even when we attempt to resist such logics. Arts-based representations of our bodies may then reflect stylized "routines" of gender, race, sexuality or other categories (Butler, 1990).

So arts-based research can be emancipatory and highlight marginalized voices, but if it is to do so, then researchers must go beyond documenting oppression by describing particular instances of it and their emotional and material consequences. Such narratives may fail to point past oppression (i.e., knowledge of subjugation) to resistance and possibilities for living otherwise. Arts-based research may embody particular knowledge of the world available only to those who occupy intersections of identities well outside dominant power structures (i.e., subjugated knowledges) or who dare to perform a "traitorous identity" and reveal the material, emotional, and cognitive manifestations of intersectional privilege (e.g., Sholock, 2012). Likewise, qualitative studies of marginalized groups, health disparities, and under-resourced communities can also intend to offer subjugated knowledges but instead merely reinforce stereotypes based upon embodied differences, such as skin color or biological sex.

On the other hand, subjugated knowledges and knowledge of subjugations are, if not a continuum, at least not discrete or dichotomous. Some stories may only hint at subversive experience and knowledges while others may be more explicit. It may be particularly difficult to understand embodied experiences without inadvertently essentializing a single identity, such as looking at the devastating effects of homophobia within families without considering its mediation through wealth, white privilege, or educational attainment. Focusing on intersectional embodiment may assist in revealing subjugated knowledges rather than replicating hierarchies. Drawing on Crenshaw (1989, 1991) and Mirza (2009), Ntelioglou (2015) suggests "embodied interactional analysis" that illuminates "lived experience of the body, taking into account their multiple social locations, analyzing how [participants'] experiences have been influenced by factors such as social, economic, and cultural contexts" (p. 91). For example, in a study of first generation college student women who went on to earn a Ph.D. and became professors, the authors describe two of their participants, highlighting intersectionality of identity and key lived experiences:

> Cassandra is Black, a full professor, near retirement, physically disabled due to mobility problems, and works in the social sciences. Sera is White, an early-career assistant professor, a single mother, and works in the humanities. Though they grew up in opposite geographical and cultural ends of the United States, and had very different childhood experiences, their trajectories to academia are quite similar.
>
> *(Jackson & Mazzei, 2013, pp. 263–264)*

Jackson and Mazzei's focus on multiplicity and ambiguity in the women's stories, and the researchers' refusal to compare these two women as though they were static, decontextualized data points, demonstrates possibilities for nuanced analysis that can highlight subjugated knowledges. Feminist researchers may also be familiar with the admonition to always continue "working the hyphens" (Fine, 1994) between self as researcher and participants as intersectional Others, complexifying identities. Hoel (2013) offers further insights into the process of uncovering subjugated knowledges by building on Haraway's (1988) concept of accountability, or the researchers' commitment to not reinforcing existing power structures, such as gender binaries, and to make explicit how power weaves throughout the situated knowledges. No matter what type of representation researchers produce, we should strive to illuminate subjugated knowledges without reinforcing current power dynamics by revealing only knowledge of how subjugation is experienced that offers no "lines of flight," or possibilities for either knowledge of the world produced from a subject position or creative resistance born of bodies immersed in intersectional oppressions.

Radical Specificity

Another generative concept that relates to subjugated knowledges is radical specificity. Drawing upon Deleuze (1994), Sotirin advocates for engagement of radical specificity in representation (Sotirin, 2010). While she addresses autoethnographic narratives specifically, I borrow one facet of her central argument to support an articulation of radical specificity and embodiment as integral to practices of representation in any genre of qualitative research. Resisting the framing of difference as variations of a single monolithic phenomenon (motherhood, in Sotirin's case), difference can be honored through an articulation of the "*radical specificity* of living a life, not in the sense that we all live our own lives but in the sense that life is lived in the flows, multiplicities, and provisionality of each moment, event, emotion" (Sotirin, 2010, n.p., sect. 6). In this way, the goal of scholarship shifts from evoking recognition, empathy, and commonality of experience to embracing "the opportunity to think beyond the dominant, the familiar, and the common" (sect. 7) to look at moments of a lived life. Sotirin urges readers to accept that there is no essential core experience which researchers and participants all express in somewhat differently embodied ways. Instead, "radical specificity opens unfamiliar connections and relations that move both beyond and against the familiar storylines, emotional verities, and the all-too-recognizable critiques of cultural-political constraints that characterize personal narratives in both popular and academic writing" (sect. 7).

In other words, radical specificity is reflected in stories that do not conform to cultural clichés—e.g., the harried working mother or the tireless warrior mother advocating for resources for her child with learning disabilities. In this way, radical specificity as a practice "creatively dismantl[es] the affective relations defining the institution and experience of motherhood [for example] and allow[s] the singularity of those relations to show us something different" (sect. 7). That "something different" can be radically specific ways of living bodies in moments, rather than generalizable truths or commonalties. For example, Lord (2004) created an assemblage of her experience with breast cancer, as a white lesbian in her 50s, a mix of emails, rants, lists of biomedical facts, and photographs—none of which fit neatly together—forming an example of queering cancer (Bryson & Stacey, 2013) through "an autobiographical account of life with cancer that explicitly and elegantly refuses the canonical requirements of biography" (Bryson & Stacey, 2013, p. 204). In this way, Lord does not provide readers ready access to empathy on the basis of their commonality of experience; rather she offers snarky comments, refers to herself in the third person as "Her Baldness," and otherwise remains irreverent. The impulse to empathy represses all that we differ on and erases points of difference in favor of emphasizing similarity. In contrast, embracing radical specificity "may not be about communicating a shared experience; rather, we might think more in terms of a 'rhizomatic' movement of senses and perceptions" (Sotirin, 2010, sect. 6).

Radical specificity applies to narratives and the opportunity to reach not for the canonical moment of commonality but for the odd, irreverent, embarrassing, or confusing moment instead. Yet radical specificity also may help to enrich the construction of themes or categories across a data set. That is, instead of gathering interview quotes and fieldwork excerpts that form variations within a coherent theme, researchers could instead (or also) think of the ways in which the radical specificity of lived moments manifests not a singular experience or identity but together those moments illuminate the intersectional complexity of lived experiences of body-selves. As an example, I want to revisit a study my collaborator and I conducted that addressed the satisfaction women with breast cancer felt with their physicians' ways of communicating (Ellingson & Buzzanell, 1999). In retrospect, we missed the opportunity to highlight the radical specificity of some moments in our data, such as the embodied experience of surgery for one participant who was pregnant when diagnosed with breast cancer. We acknowledged the particularity of participants' lives through a table in the methods section that provided bits of information about each participant's age, diagnosis, and circumstances, yet we still positioned concepts such as respect or caring as common experiences of a singular phenomenon for which we provided illustrative examples. Radical specificity reminds researchers that life is lived at the intersection of common stories with the specific embodied moments in the ebbs and flows of a particular life.

Refracting Bodies through Crystallization

Another meaningful way to do embodiment in qualitative research representations is to use a crystallization framework (Ellingson, 2009a). Richardson (2000) offered the crystal as an alternative metaphor to the positivist image of methodological triangulation as the basis for methodological rigor and validity. I further articulated this concept into a framework for qualitative research.

> Crystallization combines multiple forms of analysis and multiple genres of representation into a coherent text or series of related texts, building a rich and openly partial account of a phenomenon that problematizes its own construction, highlights researchers' vulnerabilities and positionality, makes claims about socially constructed meanings, and reveals the indeterminacy of knowledge claims even as it makes them.
>
> (Ellingson, 2009a, p. 4)

Crystallization involves engaging multiple forms of analysis and multiple genres/media of representation within a research project, making *and* destabilizing knowledge claims, yielding a postmodern validity (see also Janesick, 2000). For example, in the chapter's opening narrative I described a project that involved systematic qualitative analysis informed by feminist theory, an art installation, and a website with information intended for public consumption; together, these different representations of data position bodies in a variety of ways. Embodied knowledge production and representation is evident in crystallized research projects in at least four ways. First, varying aspects of the body become evident in different genres/media—photos, narrative, arts. Second, crystallization reveals the body as an absent presence in many genres, particularly conventional research reports. Scholars have decried the ways in which the absence of bodies in research promotes a fiction of disembodied knowledge production (e.g., Ellingson, 2005, 2006). Yet juxtaposition of narrative, poetic, visual art representations with research reports reveals the cracks in the glossy, cerebral surfaces of academic prose, making the absence present. Third, the processes of knowledge production, the ways knowledge is produced by and through a body rather than a disembodied voice, moves beyond just representation to consider the knowledge itself, rooted in carnal experience. That is, multigenre work reveals the body not as an object of cognition to be transparently or accurately represented but as a source of knowledge and a way of knowing that produces representations. Fourth, crystallization aids in resisting embodied binaries of gender, race, ability, age, and so on by revealing and challenging false dichotomies of identity. One ethnographer who embraced multigenre storytelling and analysis in his text about pugilism (boxing) explained his justification for combining narrative with analytical writing to convey a deep sense of embodiment:

> How to go from the guts to the intellect, from the comprehension of the flesh to the knowledge of the text? . . . To restitute the carnal dimension of ordinary existence and the bodily anchoring of the practical knowledge constitutive of pugilism—but also of every practice, even the least "bodily" in appearance—requires indeed a complete overhaul of our way of writing social science. . . . so as to capture and convey "the taste and ache of action" to the reader.
>
> *(Wacquant, 2009, pp. 122–123)*

Wacquant embraced a nonlinear, multigenre account to illustrate his fascinating study of pugilism. Another great example of crystallization is the work of Harris (2009, 2012, 2015) who investigated drug use and living with hepatitis C. She made a video that complemented scholarly articles, shared her personal history of drug abuse and recovery, and engaged in praxis with strategies and outreach for harm reduction in the community of addicts and recovering addicts. Crystallization offers one path to representing bodies as refracted through a prism of multimethod/multigenre analysis and representation, illuminating both material and symbolic needs of a variety of stakeholders implicated in qualitative research projects.

Practices for Embodying Representation

No single way to embody qualitative research representations has emerged as the most effective or aesthetic; qualitative researchers doing embodiment need a large toolbox from which to draw. Remember that "no single genre or method can capture it all; that nothing we say can be nested in the entirely new; and that the field of experience and representation is by definition both cluttered and incomplete" (Brady, 2004, p. 633). I urge researchers to explore a wide range of embodied representations for qualitative research data and analyses. The framing questions and the data that are co-constructed should lead researchers to their representational forms. The desire to explore a particular genre can also drive representation, although researchers should remain open to the possibility that some forms simply will not suit some data. Moreover, the experience of composing representations in any form will do things to the data through intra-action with researchers' body-selves. Composing narratives will affect the themes and categories you construct, for example, and reading a "hot spot" (MacLure, 2013) through a theoretical lens will impact your poetry and performances in a fertile cross-pollination of genres and analyses. The process of constructing representations continually changes the researcher *and* the data. The strategies below include creative ways to do embodiment in and through a variety of different genres across the continuum of qualitative methods.

Explore Intersectionality through Visual Media

Visual media offer creative means for illuminating the complexity of intersectional identities. Education researchers who embraced new materialism wanted an engaging way to teach about intersectionality and bodies to education students; they proposed to "make sense of and produce the body in graphica scholarship, hoping to transcend histories of constraint and oppression" (Jones & Woglom, 2015, p. 116). The authors created a graphic novel that shows bodies, positioned and with facial expressions, in specific contexts and highlighted intersections of race, gender, class, and age, bringing materialism into conversation with competing discourses of racism, sexism, heteronormativity, and sexual violence (see also the drug prevention comic developed by Miller-Day, Conway, & Hecht, 2013). Other examples of visual arts that explore theory and research results include Fenge and Jones (2012), who conducted a participatory research project about the experiences of aging lesbians and gay men in rural areas and produced a short film to share participants' lives with wider audiences. Other great examples of embodied visual representation include Harter and Hayward's

(2010) documentary which features five families coping with a child's cancer treatment, Harter, Shaw, Ruhl, and Hodson's (2015) documentary on the Arts in Medicine program at MD Anderson's Children's Cancer Hospital, and Harter, Shaw, and Quinlan's (2016) documentary on a sheltered workshop for adults with developmental disabilities that integrates art programming (see also Ball and Gilligan's [2010] discussion of social science-visual arts collaborations to explore migration and social divisions).

Body-Becoming Pedagogy

Resisting instrumental outcomes, a body-becoming approach focuses on creativity and possibilities for bodies that are deemed by cultural standards to be different. Rice (2015), a fat studies theorist, drew on Grosz's (1994) corporeal feminist theory and proposed a theory of fat "informed by feminist, poststructuralist, and new materialist bioethics [that] would move away from cultural practices of enforcing norms toward more creative endeavors of exploring physical abilities and possibilities unique to different bodies" (p. 392). Her theorizing is specific to understanding fat bodies in a society that pathologizes bigger bodies without regard to the genetic, cultural, and structural forces that produce and are produced by fat bodies. However, her theory has additional relevance for doing embodiment in qualitative research. She wrote:

> body-becoming pedagogies—as interventions that create alternatives to conventional biopedagogies whose instrumental, outcome-oriented methods and moralizing overtones enforce physical conformity over diversity and creativity. Rather than being instrumental and outcome-driven, a body-becoming pedagogy is presence- and process-oriented.
>
> (Rice, 2015, p. 395)

I embrace such an approach as applied to harnessing the potentials of *all* bodies, but most especially marked bodies excluded on the basis of differences from an idealized norm, including but not limited to disabled bodies, bodies of color, undocumented immigrant bodies, bodies in poverty, genderqueer and nonheteronormative bodies, transbodies, pregnant bodies, menstruating bodies, aging bodies, sick bodies, bodies without homes. Such a pedagogy, or learning practice, invites us to learn from, about, and through bodies of all types, appreciating commonalties and differences through a playful, process- and pleasure-oriented aesthetics that focuses on possibility and imagination rather than shame, lack, and conformity. When doing embodiment through your research processes, connect with the concept of a body-becoming pedagogy and become an appreciative student of intersectional bodily differences. Ask how all bodies in your research move through the world rather than disparaging or pitying ways of movement that seem limited or constrained, or describe a range of skin colors, features, and textures rather than reducing people to white and not. Resist bodily dichotomies as much as possible and embrace possibility and imagination to embrace bodies in all their variety as valuable and deserving of dignity. While this implication certainly relates to all phases of research, the point of representation is central to Rice's position, who examined art featuring fat bodies in order to frame an appreciative aesthetics; a similar case can be made for representational aesthetics that honor bodies of all shapes, sizes, and configurations.

Mind-Body Your Metaphors

A strategy for embodying analytic accounts is to use metaphoric language in developing results and in all forms of representation. Researchers "must choose our words carefully and creatively, with attention to the consequences of naming experience" (DeVault, 1990, p. 110); our choices are political and form part of the scaffolding of social structures, maintaining (and resisting) status quo power relations. The language we use to describe the results and outcomes of our research—themes, categories, types—influences what we share with others. Metaphor "not only provides a shortcut, but reinforces particular mental images, framing experience in a particular, political, and powerful way. They can literally make us see our world differently" (Markham, 2013, sect.2, ¶7; see also Paterson, 2009). Likewise, "[m]etaphors are slices of truth; they are evidence of the human ability to see the universe as a coherent organism" (Danesi, 1999, p. 111). For example, researchers can use their embodied responses to data, such as a fieldnote or transcription segment, "as an aid to the imaginative conceptualization of categories" (Rennie & Fergus, 2006, p. 491); felt emotions and associated physical metaphors (e.g., feeling *crushed* when grieving, as though one is *falling* in love, *dizzy* with joy, *wounded* by betrayal, *struck* by a sharp contrast) are sources of rich, nuanced names for conceptualizing and describing categories. For example Mailman (2012) offers seven metaphors for listening that challenge readers to reconceptualize listening in creative and provocative ways, including digestion, recording, adaptation, meditation, transport, improvisation, and computation (see also El Refaie, 2014 for a discussion of dynamic embodiment, metaphor, and cancer). Metaphors can inform performances of data and ethnographic performance as well (Myers & Alexander, 2010). An outstanding example is exploring the experience of living with inflammatory bowl disease via the metaphor of "coming out of the [water] closet," explored by Defenbaugh in her performances and performative writing (e.g., Defenbaugh, 2011, 2013). Harness the tool of metaphor for representations across the continuum from art to science, selecting those metaphors that enhance your ability to embody accounts.

Change the Flow of Text

Push the envelope for how journals typeset articles to reflect embodied knowing. Turner and Norwood (2013) reflected on gendered embodiment by having the three narrative sections of their analysis organized as nested narratives, with one story in a shaded box in the center of the page, and the other story wrapping the text across the top, along both sides, and across the bottom. Lather and Smithies, (1997) book on women living with HIV/AIDS presented dual narratives with participants' dialogue on the top half of each page (visually privileged, placed above, the women labeled "co-researchers") and the analysis running along the bottom of each page. Another format features three stories running as parallel accounts of abuse, including an offender in the left column, a victim in the right column, and their accounts periodically interspersed with the researcher's thoughts in a middle column (Fox, 1996).

Destabilized Narratives

If you are planning to write ethnographic or autoethnographic narratives, consider a "destabilized narrative" (Frank, 2000) in which

> different points of view are shown, but not assimilated into the authorial voice. This type of narrative is more open-ended and demands more from the reader. It does not do all of the thinking for the reader. Instead, it shows how messy and contingent reality can be.
>
> *(Richards, 2008, p. 1723)*

Typically, narrative coherence is a hallmark of a "successful" ordering and interpretation of events into a story (Harter, 2013b). Yet when narratives smooth out the wrinkles and bumps of reality too far, they can close off multiple layers of sense making and the ways in which narratives inhabit different bodies in a myriad of ways. Rambo Ronai's (1995) classic layered account of child sexual abuse is still one of the best examples of a deeply embodied, fragmented, layered narrative that illustrates connections among concepts, bodies, and actions while still pointing to multiple points of disconnection, ambiguity, and complexity. It is exquisitely painful to read the details of her little girl's body's resistance, resignation, and resilience. Jago's (2002) account of depression, Day's (2010) performative text about family stories of sexual abuse, and Minge's (2007) fragmented story on embodied memories of rape and of love resonate strongly as destabilized narratives that consider the complexities of multiple point of view. The *Handbook of Autoethnography*, edited by Holman Jones, Adams, and Ellis (2016), cites numerous other examples, ethical guidelines, strategies, and practices (see also Faulkner, 2016).

Poetry and the Body

The creative analytic practice of poetic representation of data provides a range of rich possibilities for researchers to do embodied poetic representation (Carr, 2003; Chawla, 2008; Ellingson, 2011b; Fox, Humberstone, & Dubnewick, 2014; Furman,

2006; Glesne, 1997; Hartnett, 2003; Lahman & Richard, 2014; Richardson, 1992; Willis, 2002). "Poetry has the potential to disrupt the taken for granted" (Denshire, 2014, p. 839) and is particularly well suited to writing the body, because poetry "takes its motivations and saturations from the Made World, from our dreams, and from the connections of both to the physical presence of being. It creates and occupies sensuous space that is connected to the deepest level of existence" (Brady, 2004, p. 630). Poetic writing is inherently sensual, playful, and immersed in the specific moments of specific lives; the genre itself is a refusal of objectivity. "Poetry puts a semiotic smudge on [positivism's] window, offers no free vision, [and] shows itself as method" (Brady, 2004, p. 628). Faulkner's (2009) guide offers a nuanced, theoretically grounded discussion of poetry as method and practical strategies for writing poetry in research (see also her collection of poems, Faulkner, 2014). Esposito (2014)'s poetic representations of chronic pain, embodiment, and relationships is an outstanding exemplar. I also highly recommend Sarah Kay's (2011) TED talk on spoken word poetry and encourage you to enter "spoken word poetry" into YouTube and revel in the wealth of spoken word poetry/poetry slam videos as inspiration of how poetry is inextricably tied to bodies and how deeply felt emotions shape and are shaped by the body-self.

Ethnodrama

Performance comes through the body. Saldaña (2003) suggests that ethnodrama, or dramatizing ethnographic research to tell stories through theatre, should be done when the medium fits with the story needing to be told, not just done for the sake of dramatizing.

> This may be difficult for some to accept, but theatre's primary goal is neither to "educate" nor to "enlighten." Theatre's primary goal is to entertain—to entertain ideas and to entertain for pleasure. With ethnographic performance, then, comes the responsibility to create an entertainingly informative experience for an audience, one that is aesthetically sound, intellectually rich, and emotionally evocative.
>
> *(Saldaña, 2003, p. 220)*

Saldaña (2011) offers a practical and engaging primer on developing ethnodrama and ethnotheatre productions, including generating a script from data. An example of an ethnodrama features a patient with a team of researchers. The first author is a patient who had chronic kidney disease and collaborated with research partners (Schipper, Abma, van Zadelhoff, van de Griendt, Nierse, & Widdershoven, 2010). Another ethnodrama focuses on youth in a detention center and the staff who work with them (Conrad, 2012). Ethnodrama often works well as a participatory research project (see Miller-Day, 2008; for an anthology of ethnodramas, see Saldaña, 2005).

Link Website or Blog to a Publication

Researchers' reflections, stories, rich details, and further examples from their data often cannot fit within the strict length parameters maintained by most journals. However, some journals now provide "supplemental materials" areas on their websites where authors can post further illustrations of their work, providing more opportunities to address embodiment directly through reflections and to offer descriptions and stories imbued with sensory details. Researchers can also create and maintain their own blogs or websites. One group of collaborating researchers created a wordpress.com blog to allow readers of their article the option of learning more about who they are through narratives and poetry that describe their history of experiencing their bodies in movement, including a statement that invited readers to learn more about their body-selves: for additional information "about the six collaborators please see our biographies, particularly as they relate to our academic-active selves, available at: www. movedtomessiness. wordpress.com" (Avner, Bridel, Eales, Glenn, Walker, & Peers, 2014, p. 56n2). Social media sites also provide opportunities for sharing embodied details and images, such as Tumblr and Facebook.

Go Guerilla

Guerilla scholarship, as I have (re)conceptualized it (Ellingson, 2009a), based on Rawlins' (2007) articulation, can take a multitude of forms, with the goal of sneaking disruptive perspectives, genres, meanings, and bodies into places they do not fit or are not welcome. Two primary strategies of guerilla scholarship are to pass and to claim public audiences. Passing is about fitting in publishing conventions while sowing seeds of embodied subversion of dominant paradigms. Some articles may have to conform to immediately postpositivist conventions that may not fit comfortably with poststructuralist, feminist, or narrative approaches to research, but which remain the normative form for many professional journals. But there are alternatives and opportunities to subvert the norms to at least hint at embodiment while conforming to them sufficiently to gain entrance. Rawlins (2007) explains:

> I was still seduced in all my earlier work by the dominating ethos of quantitative social science into aping its trappings, writing style, and subdivisions . . . in order to *pass* as a serious researcher. I call such activity *guerilla scholarship* . . . The stated and unstated regimes of certain journals require these kinds of accouterments.
>
> *(p. 59)*

If subterfuge is required to effect change, then do it. Hold tight to your body-self and the embodied knowledge you and your participants have produced, and approach a strategy of embodied subversion of writing norms. One strategy is to cite within an article multiple other works that reflect different methods, genres of representation, ideologies, even paradigms. Reference sections fulfill this function, of course, but footnotes, epigraphs, highlighted quotes in an essay or report, and interludes or other interrupting discourses can all be places to invoke the methodological or representational Other. Methods sections also offer good places to highlight embodied standpoints. You can also push the envelope through the occasional provocative or creative embodied phrase that challenges the standards of acceptability in a publication outlet and thus subversively broaden the horizons of the publication.

A second form of guerilla strategy is doing research dissemination to multiple publics who can benefit from the ideas and strategies that you have developed (Koro-Ljungberg & MacLure, 2013). Researching diverse publics involves translating and re-visioning academic representations to ones suitable for other audiences, such as newsletter articles, website content, editorials in online and print news outlets, plays and performances, documentaries, and workshops or skills trainings for professionals and for community members (Ellingson & Quinlan, 2012). These forms of public outreach are often referred to as "public intellectualism" but we can also think of them as opportunities for embodying the intellect by highlighting sensory details, intersectional identities, and embodied knowing as we reinvent our scholarship for wider audiences. Undoubtedly there are a multitude of other ways to do the body subversively in venues where it is an unwelcome presence; explore and experiment with how to best be sneaky. Of course, we can also publish in venues where embodied narratives and embodied knowing are celebrated, and I urge you to do so. But I also contend that chipping away at the vestiges of positivism and at the Cartesian banishment of the body that constrain the representational norms of "mainstream" journals is a vital enterprise.

Note

1 This exciting, complex, and varied body of theory has exerted significant influence on qualitative and critical methodologies (e.g., Vannini, 2015), in a manner overlapping significantly with poststructuralism, actor network theory, posthumanism, and other contemporary theorizing of embodiment, including the concepts that I introduced in Chapter 1. I fully acknowledge the partiality and partisan nature of all research accounts, the need to make space for flux and becomings rather than conclusions and static findings, and the benefits of bridging art and science to create hybrid, multi-sensorial accounts. Clearly, dedicated non-representational theorists would object to many aspects of this theoretically promiscuous (Childers, 2014) book, up to and including my use of the term representation.

POSTSCRIPT

Common Threads

Doing Legwork: A Calling

I watched anxiously as the last of the passengers left the gate area, proceeding down the broad airport hallway in response to the gate change announcement. An American Airlines representative gave a distracted smile to me and the man next to me, calling out as she left, "Someone will be back to get you." I sighed.

"Hope they send someone soon," commented my companion. He and I were stranded in airport wheelchairs, their stainless steel parts shining in the harsh fluorescent lights.

Only three months out from the above-knee amputation of my right leg, I wasn't strong enough to walk to the distant gate on my crutches. "Hope so. I need to make this flight, for sure," I responded, turning to smile at the gentleman. "I'm going to a conference." A fringe of white hair ringed a mostly bald head marked with several brown age spots. His watery, pale blue eyes were alert and focused on me.

As we waited, the man kept glancing at my simple prosthesis, the precursor to the computerized one I would eventually receive. "Were you in the military?" he asked, looking at my leg again.

"Ah no. I had bone cancer in my thigh years ago, and the doctors rebuilt it with bone grafts and metal implants. Then a series of staph infections led to the amputation this past summer," I explained patiently, having answered this question several times already on this trip to other inquisitive travelers and flight attendants.

Pain etched itself on the man's face. "The bastard took my wife!" I nodded, not sure what he meant by that. He continued, "She had lymphoma three times, and she fought real hard. She beat it twice. And then the third time she, she . . . She just . . . " he trailed off, stricken. Ah, the bastard was cancer.

"I'm so sorry about your wife," I said automatically. "That's so sad. It's so . . . hard," I finished weakly.

His eyes delved into mine, seeking understanding. "The third time she didn't want to fight again. She told me she was too tired, that it was time to go." Anguish twisted his face. "The bastard!"

I smiled sympathetically. "I'm so sorry. I'm grateful to survive," I said. "And I never understand why some survive and others don't, can't, that is, I mean, they aren't able to." My face flushed as I stumbled.

"Yeah," he said, nodding, still looking into my eyes. "Yeah."

"It's good that she had you to take care of her," I offered hesitantly.

"Mmm," he responded, nodding. Our gazes held for a few moments, and then he looked off into the distance, seemingly lost in thought. A younger man strode with purpose toward my companion and released the brakes on his wheelchair without addressing the seated man. With no choice but to go where pushed, my companion headed out. "I won't let them leave without you!" he called, smiling at me and waving his hand briefly.

"Thanks! Nice to meet you. Hope you have a good flight," I responded. Soon after, another efficient airline employee arrived to push me to the gate, and I pondered my encounter with the older man who needed to be heard.

I am less a storyteller than a story gatherer. My body tells (some of the) stories about the way my body-self lives in the world, and these embodied narratives spark other people's bodies to call out to me for acknowledgement and connection. This has been true throughout the 28-year saga involving my diagnosis, treatment, and recovery from bone cancer in my right leg, as well as my affliction with "late effects" of cancer treatment, i.e., all the ways in which your body pays you back for enduring the chemotherapy, radiation, medications, surgeries, and other stuff the medical experts subject you to while trying to save your life.

Later on the airplane, I thought with compassion of the man who poured his heart out to me at the gate, and I recalled a comment my husband Glenn had made to me several years prior when I was still undergoing reconstructive surgeries on my right leg. We had been walking into the Old Meeting House, a popular local hangout, with our friend Elena. As I lurched gracelessly forward on crutches with a long brace with Velcro straps covering my leg, I was stopped by a seated customer who asked what had happened to my leg, an everyday occurrence for me in the friendly Southeastern region of the U.S. where I attended graduate school.

After I answered her politely and offered sympathy about her recent knee replacement surgery, I joined Glenn and Elena who were already seated at our booth. Well used to such interruptions, Glenn looked at me, a warm smile lighting his face, and quipped, "You could do a paper called 'Big Blue Brace as Method!'"

Glenn's comment was made several years ago, and now as a recent amputee, I couldn't help but wonder how my body would speak to others, especially strangers, and how the stories my body tells would differ following this last, definitive surgery that ended the quest to repair my faulty limb. Instead of a big, blue brace, I have a computerized prosthesis that has other stories to share. My brief interaction with the passenger who had recently lost his wife to cancer intrigued me as much as it saddened me to hear his story. I also felt a sense of gratitude that I had been able to bear witness to his pain; my body-self—marked with pain and suffering—offered him a moment of connection and empathy across our many differences, and I was glad.

Pulling Threads

As the above narrative illustrates, I feel called to the work of embodiment—in both senses of the term. People see my body (when I am conducting research and in my everyday life) and call out to me, usually wanting my body stories and sometimes the chance to share some of their own. I also feel a deep calling, or vocation, to conduct feminist, qualitative research that addresses marginalized topics and people and tries to foster material change toward a more just and humane world. I particularly enjoy stepping back from my practice of feminist, qualitative research to write about possibilities for the practice of research.

I want to wrap up this book by reflecting on several threads, or themes, that run through the chapters. Novelist Anita Shreve (2010) said, "A novel is a collision of ideas. Three or four threads may be floating around in the writer's consciousness, and at a single moment in time, these ideas collide and produce a novel." That is pretty much what happened to me; over time, several threads about embodiment theorizing and qualitative research collided to form this book. The threads wove organically across my writing, even as I organized the book in a fairly conventional manner, with each chapter addressing a recognizable topic area in qualitative methodology. I briefly discuss the threads here as a way of both concluding the book and of opening it up further to becoming what readers make of it as they grasp one or more of these threads or hold onto threads that I (at this time) have not recognized.

The first thread is troubling the taken-for-grantedness of basic details that form routine practices within qualitative research. I have endeavored to reflect on how bodies are reduced to containers for selves located in brains, taken as transparent, or dismissed as inconsequential in many method practices. From selecting participants whose bodies reflect specific categories or have undergone particular experiences, to observing bodies at work and at play, to recording and transcribing the speech of bodies, to constructing narratives and poetry of bodies in relationship with others, qualitative researchers simultaneously invoke and ignore the material, embodied nature of people and render knowledge as something generated by minds in spite of, or at the impetus of, but seldom through, our bodies. Reconsidering the smallest details of methods practice in terms of how they implicate researchers' and participants' body-selves, actants, and places/contexts is productive for all phases of the research processes. Across the qualitative continuum, researchers tend to specialize in a relatively small set of practices—autoethnography or grounded theory, mixed method designs that use surveys and interviews or critical readings of interviews and observations through a theorizing lens—and we come to mentally take for granted the smallest details of practice. Recall, as I addressed in Chapter 6, that even the ability to cut-and-paste blocks of text on a computer is a material practice that necessitates a body positioned to use a computer, hands, or voice commands, eyes to see the text or ears to hear it, and lungs to breathe while the bodily task is accomplished. Rethinking basic research practices as material opens up opportunities for seeing them from a fresh perspective, capitalizing on their strengths, and imagining (equally embodied) alternatives.

A second thread is creativity; research is the process of creating data, creating analyses, creating stories and performances, creating knowledge, and (directly and indirectly) creating social change. Bodies link with minds and spirits in the creative process. Those at home on the artsy end of the qualitative continuum may already embrace the mind–body connection in their process and hopefully have gained some further nuances and possibilities for their creative practices from this book. For those in the middle and science areas of the continuum, the invocation of research as embodied creativity is likely new, probably odd, possibly disturbing or even threatening, and hopefully intriguing and exciting. Understanding research as embodied creativity—making things with our bodies—opens up possibilities for understanding research practices not just as abstractions (although abstractions have much to offer, too) but as material actions with material consequences.

Blaspheming the sacred within qualitative methodology is another important thread in exploring embodiment. I embrace a continuum of methods, intersected with critical lines of flight, and reject the thick lines drawn between paradigms. Authentic, emotional voices can ring through autoethnographic narratives *and* through grounded theory analyses. Rigorous analyses can be conducted by (post) positivist, interpretivist, *and* artistic methods. The body constitutes irrefutable proof that objectivity—Haraway's (1988) "god trick," the view from nowhere, positivism's standard for validity, and postpositivism's regulatory ideal—is an impossibility. And it also demonstrates that no researcher, using any method, escapes her or his intersectional standpoint or captures Big-T-Truth. My goal in connecting embodiment theorizing with qualitative research practices has been to demonstrate that those in the middle and science areas of the continuum can do embodiment in their research *without* having to abandon their usual practices and take a giant step toward art on the continuum, and that those in the arts and post-qualitative areas can consider bodies from wider perspectives that include insights from interpretivism and postpositivism.

While I encourage all researchers to venture outside of their accustomed areas of the continuum to try new methods and forms of representation from time to time, I do not argue for the innate superiority of some forms of research, including in the service of embodiment. Yes, live performances invigorate through their embodied manifestation of emotion and movement, and narratives illuminate bodies engaged in dialogue, relationships, and action. But interpretive, critical, and structured analyses offer vital insights into bodies and power, common threads that connect bodies, cultural constructions (and negations) of bodily value, and people's sense making about their bodily experiences of suffering, striving, and thriving. Different methods encourage researchers to do embodiment differently, and each offers opportunities and constraints. I do not deny differences between paradigms, but I urge researchers not to fetishize them either. All paradigms are embodied human constructions; we made all of these rules up, and we can play within them and outside of them.

On a similar thread, I not only rejected paradigmatic boundaries, but also refused those between bodies of embodiment theorizing as well. After reading thousands of pages of feminist, poststructuralist, postmodern, new materialist, posthumanist, postcolonialist, and other forms of theorizing, and writing and reflecting over time, I selected the seven concepts highlighted in Chapter 1 that weave throughout the subsequent chapters. As far as I know, this choice struck some colleagues as innovative and others as objectionable; undoubtedly still others declined to share with me differing evaluations of this strategic choice. As with any choice, much was lost; I largely ignored important and meaningful differences and even incompatibilities between bodies of theorizing, and I completely ignored many generative concepts. I included the concepts that made my mouth and throat hum, my eyes widen, my spine straighten, my head nod, and my shoulders shiver with excitement when I read about them. I chose exemplars that made my heart speed up and my breath gasp. My embodied responses guided the decisions, and while I stand by their usefulness, I also readily acknowledge that they are eclectic choices based on what appealed to my body-self, mostly during the last three years when I was very focused on this project, but also since my cancer-survivor-body-self began my first study of women with cancer in 1996.

A relentless thread of pragmatism also runs through this text. I am a big proponent of practicality; embrace whatever (ethical) strategies will get the job done, however you define the job. Cross paradigms, work with community partners, join an interdisciplinary research team, change your approach to data collection because your participants offer unexpected access or refuse to cooperate with guidelines you thought you had all agreed upon. One of the beauties of qualitative methodology is its flexibility and practicality; projects grow and change over time, analyses are iterative, participants depart and others join, grant money ebbs and flows. When I could not stand up during an extended observation period (due to leg pain and fatigue), I embraced a whole new perspective on the dialysis unit from a seat at the nurse's station, and that perspective was equally generative to the one rooted in more standing and walking around. Embracing the flux of our research participants and settings, as well as the flux of our own everyday lives as they intersect with our research practices. Is not haphazard, nor is it lacking in rigor; rather, it reflects a creative, generative commitment to embodied pragmatism.

Materializing Social Change

My final thoughts are a further invitation to do embodiment in qualitative research in ways that will foster compassionate change in the world, affecting the material conditions of bodies. A false dichotomy persists between academia with its analysis, reasoning, and artwork on the one hand, and the functioning of the rest of society on the other. Abstract theorizing is not irrelevant to teaching, loving, cleaning, and community building. Researchers' body-selves can be a bridge between the academy and the rest of the world, and we can help to make material differences in the world in many ways, all of which we can support by doing embodiment in our research practices. "Confronted with personal suffering and the overwhelming inequity that characterizes many societies, it is easy to doubt the capacity of research to make any real difference" in the world (Warr, 2004, p. 579). Yet we can persist in reaching out to our communities to share what we are learning. For example, UNC-Charlotte researchers Margaret Quinlan and Bethany Johnson shared their expertise and insights from their study of infertility on a local radio program, which was a small but important way to offer information and compassion to people struggling with infertility.

We can believe in possibility while still taking concrete steps. I want to share two specific, useful skills researchers can develop through doing embodiment.

First, researchers can develop improved empathy as a skill. "Being present in the telling of these uncomfortable stories can heighten the capacity of the researcher to portray people's experiences with empathy and a deeper level of understanding" (Warr, 2004, p. 581). Increasing our capacities for empathy is a powerful strategy for improving our ethnographic practices and products and for helping to make our worlds better places. Empathy and compassion are linked as embodied practices, and therefore skills that can be learned, practiced, and strengthened, along with appropriate and respectful boundaries. A communicative model of compassion centers on the embodied ways in which compassion is expressed and understood, and the model divides the practice into three components (Way & Tracy, 2012):

- *recognizing* a person's needs by actively listening and attending to both verbal (what is said) and nonverbal cues (e.g., body language, facial expressions) and how those two types of cues complement, complicate, or even contradict each other
- *relating* to another by "identifying with, feeling for, and connecting with another in a consubstantial way" (p. 303)
- *(re)acting* to a person by taking compassionate actions; "the parentheses around 're' indicate that compassionate action need not only be in response to or arise after the recognition of someone else in pain, but can also be proactive" (p. 307).

FIGURE PS.1 Researchers Quinlan and Johnson Participate in Public Radio
Program.
Bethany Johnson and Margaret M. Quinlan, UNC Charlotte, discuss
their research on infertility treatment support as panelists on the
Charlotte Talks program at local NPR station, WFAE.

Courtesy of Bethany Johnson and Margaret M. Quinlan (participants), and Jennifer Worsham
(photographer).

Breaking the process down into steps may help frame compassion and empathy as processes rather than simply as feelings or intuition. In this model, recognizing takes careful observation and willingness to pay attention to another's expressed emotions and needs, which then prepares you to relate, or to identify with the other person, and then to take action or be proactive in assisting another. For researchers, that action may be as simple and as difficult as bearing witness to pain, staying present with someone in their difficulties, and trying to set aside your own life experiences and embrace another's perspective on a complex problem. Empathizing comes more easily to some than others, yet the skill of compassionate perspective taking is one that can be learned through a willingness to consciously make an effort to communicate compassion and empathy and to practice again and again. This skill will benefit qualitative researchers in understanding and connecting with their participants in meaningful ways, enhancing both the richness of the research and opportunities to foster social justice.

A second specific skill set for promoting embodied social justice is the challenging work of helping to bridge multiple communities through coalition building (Reagon, 1983). Feminists have long advocated political action "through coalition—affinity, not identity," when activists work to "build unities, rather than to naturalize them" (Haraway, 1991, pp. 155, 158); that is, coming together on the basis not of who we *are*—an assumption of common (embodied) identity—but of what we *want*—a common desire to foster equity and social justice. Through shared values and purpose, researchers can come together with nonprofit organizations, community groups, government agencies, and even for-profit companies to promote positive social change.

The transformative research paradigm offers some good ways of thinking about embodied coalition building and partnerships (Mertens, 2007, 2010, 2012). Mertens proposes that culturally appropriate, mixed methods (quantitative and qualitative) research can help promote social change that is responsive to race, gender, and disability, among other embodied differences that manifest in social inequities. Fundamental to this paradigm is the stance that not all perspectives on injustice and disparities are equally valid; bodily privilege limits what is detectable to members of different groups, and subjugated knowledges offer critical insights into addressing material inequities. Mertens does not limit the types of data collection, analyses, or representations that can be used; instead she argues for sensitivity and building trusting relationships with communities in order to conduct responsible research that sparks and supports social change. Other models for embodied coalition building can be found in indigenous methods, which focus on ethical and practical ways to conduct participatory research with members of indigenous groups that honor participants' ways of knowing and of engaging with/as scholars (e.g., Kovach, 2015; Simonds & Christopher, 2013; Wilson, 2008).

Knot

Throughout this book, I have tried to articulate a way of doing embodiment that encompasses deliberate attention to material reality and to the intersection and intra-action of participants' and researchers' dynamic body-selves with cultural discourses, material objects, and everyday life. Doing embodiment is active, deliberate, conscious, but also in flux, ever becoming, the result of mistakes and accidents as much as plans and choices. The threads of my passion for embodiment weave together narrative, theorizing, research, and practices that offer many different points of entry for researchers across the qualitative and mixed methods continuum who are intrigued by the wealth of scholarly discourse on embodiment now circulating in and through us.

I invite researchers to not just think about the ideas offered here but to embrace and enact some of them, whichever ones speak to your gut, your heart, your breath. You can make space for more reflection, make a small change in your typical research practices to attend more directly to bodies, or perhaps try something wildly divergent from your usual approach to sense making. I know that my research is at its very best—as scholarship, as compassionate and empathetic engagement with the world, and as critical intervention into inequities and suffering—when my unruly, passionate, leaky, limping, loving body-self is fully present and attending with great care to the material being of others' body-selves. Whether I am doing fieldwork, interviewing participants, laboring over transcriptions, or playing with my data—surrounded by analytic memos, interview fragments, ideas for stories, bits of poetry, photos, scrawled notes and questions, my purple clipboard and pen, and my laptop spread out around me on the couch where I lounge in my pajamas, thinking, thinking, thinking as my phantom limb pain pulses and my cats purr—I am a body-self making sense with, of, and through other embodied people and our social worlds.

REFERENCES

Agarwal, V., & Buzzanell, P. M. (2015). Communicative reconstruction of resilience labor: Identity/identification in disaster-relief workers. *Journal of Applied Communication Research, 43*, 408–428. doi: 10.1080/00909882.2015.1083602

Ahmed, S. (2000). *Strange encounters: Embodied others in post-coloniality*. London, UK: Routledge.

Ahmed, S. (2002). This other and other others. *Economy and Society, 31*, 558–572. doi: 10.1080/03085140022000020689

Allen, B. J. (2010). *Difference matters: Communicating social identity* (2nd ed.). Long Grove, IL: Waveland Press.

Allen, L. (2005). Managing masculinity: Young men's identity work in focus groups. *Qualitative Research, 5*, 35–57. doi: 10.1177/1468794105048650

Arnfred, M. (2015). Polyphonic sound montages: A new approach to ethnographic representation and qualitative analysis. *Journal of Organizational Ethnography, 4*, 356–366. doi: 10.1108/JOE-10-2014-0034

Ash, J., & Gallacher, L. A. (2015). Becoming attuned: Objects, affects, and embodied methodology. In M. Perry & C. L. Medina (Eds.), *Methodologies of embodiment: Inscribing bodies in qualitative research* (pp. 69–85). New Brunswick, NJ: Routledge.

Ashmore, M., & Reed, D. (2000). Innocence and nostalgia in conversation analysis: The dynamic relations of tape and transcript. In *Forum Qualitative Sozialforschung/Forum: Qualitative Social Research, 1*(3). Retrieved from: http://www.qualitative-research.net/index.php/fqs/article/viewArticle/1020/2199

Atkins, L. (2015). Half the battle: Social support among women with cancer. *Qualitative Inquiry, 22*, 253–262. doi: 10.1177/1077800415574911

Atkinson, P. (2006). Rescuing autoethnography. *Journal of Contemporary Ethnography, 35*, 400–404.

Atkinson, P., Coffey, A., Delamont, S., Lofland, J., & Lofland, L. (Eds.). (2001). *The handbook of ethnography*. Thousand Oaks, CA: Sage.

Avner, Z., Bridel, W., Eales, L., Glenn, N., Walker, R. L., & Peers, D. (2014). Moved to messiness: Physical activity, feelings, and transdisciplinarity. *Emotion, Space and Society, 12*, 55–62. doi: 10.1016/j.emospa.2013.11.002

Bairner, A. (2011). Urban walking and the pedagogies of the street. *Sport, Education, and Society, 16*, 371–384. doi: 10.1080/13573322.2011.565968

Ball, S., & Gilligan, C. (2010). Visualising migration and social division: Insights from social sciences and the visual arts. In *Forum Qualitative Sozialforschung/Forum: Qualitative Social Research, 11*(2). Retrieved from: http://www.qualitative-research.net/index.php/fqs/issue/view/34

Ballard, D. I., & Seibold, D. R. (2004). Communication-related organizational structures and work group temporal experiences: The effects of coordination method, technology type, and feedback cycle on members' construals and enactments of time. *Communication Monographs, 71*, 1–27. doi: 10.1080/03634520410001691474

Balomenou, N., & Garrod, B. (2016). A review of participant-generated image methods in the social sciences. *Journal of Mixed Methods Research, 10*, 335–351. doi: 10.1177/1558689815581561

Barad, K. (2007). *Meeting the universe halfway: Quantum physics and the entanglement of matter and meaning.* Durham, NC: Duke University Press.

Barnacle, R. (2009). Gut instinct: The body and learning. *Educational Philosophy and Theory, 41*, 23–33. doi: 10.1111/j.1469-5812.2008.00473.x

Bates, C. (2013). Video diaries: Audio-visual research methods and the elusive body. *Visual Studies, 28*, 29–37. doi: 10.1080/1472586X.2013.765203

Baxter, L., & Hughes, C. (2004). Tongue sandwiches and bagel days: Sex, food and mind-body dualism. *Gender, Work & Organization, 11*, 363–380. doi: 10.1111/j.1468-0432.2004.00238.x

Beauboeuf-Lafontant, T. (2009). *Behind the mask of the strong Black woman: Voice and the embodiment of a costly performance.* Philadelphia, PA: Temple University Press.

Becker, L. C. (2005). Reciprocity, justice, and disability. *Ethics, 116*, 9–39. doi: 10.1086/453150

Beckers, R., van der Voordt, T., & Dewulf, G. (2016). Why do they study there? Diary research into students' learning space choices in higher education. *Higher Education Research & Development, 35*, 142–157. doi: 10.1080/07294360.2015.1123230

Belzile, J. A., & Öberg, G. (2012). Where to begin? Grappling with how to use participant interaction in focus group design. *Qualitative Research, 12*, 459–472. doi: 10.1177/1468794111433089

Berger, L., & Ellis, C. (2002). Composing autoethnographic stories. In M. V. Angrosino (Ed.), *Doing cultural anthropology: Projects for ethnographic data collection* (pp. 151–166). Prospect Heights, IL: Waveland Press.

Bernays, S., Rhodes, T., & Terzic, K. J. (2014). Embodied accounts of HIV and hope using audio diaries with interviews. *Qualitative Health Research, 24*, 629–640. doi: 10.1177/1049732314528812

Berry, K. (2016). *Bullied: Tales of torment, identity, and youth.* New York: Routledge.

Best, A. L. (2000). *Prom night: Youth, schools and popular culture.* New York: Routledge.

Best, A. L. (2003). Doing race in the context of feminist interviewing: Constructing whiteness through talk. *Qualitative Inquiry, 9*, 895–914. doi: 10.1177/107780040 3254891

Billingslea, R., & Ellingson, L. L. (2015, September–December). Voicing late effects: Stories of long-term cancer survivorship [Art installation]. Santa Clara, CA: Special Collections Gallery, Santa Clara University.

Birk, L. B. (2013). Erasure of the credible subject: An autoethnographic account of chronic pain. *Cultural Studies ↔ Critical Methodologies, 13*, 390–399. doi: 10.1177/1532708613495799

Blumer, H. (1954). What is wrong with social theory? *American Sociological Review, 18*, 3–10.

Bobel, C. & Kwan, S. (2011). Introduction. In C. Bobel & S. Kwan (Eds.), *Embodied resistance: Challenging the norms, breaking the rules* (pp. 1–10). Nashville, TN: Vanderbilt University Press.

Bochner, A. P. (2014). *Coming to narrative: A personal history of paradigm change in the human sciences.* Walnut Creek, CA: Left Coast Press.

Bodén, L. (2015). The presence of school absenteeism: Exploring methodologies for researching the material-discursive practice of school absence registration. *Cultural Studies ↔ Critical Methodologies, 15*, 192–202. doi: 10.1177/1532708614557325

Bolen, D. M., & Adams, T. E. (2017). Narrative ethics. In I. Goodson, A. Antikainen, P. Sikes, & M. Andrews (Eds.), *The Routledge international handbook on narrative and life history* (pp. 618–629). New York: Routledge.

Bordo, S. (2004). *Unbearable weight: Feminism, Western culture, and the body.* Berkeley, CA: University of California Press.

Borofka, K. G. E., Boren, J. P., & Ellingson, L. L. (2015). "Kind, sensitive, and above all honest": Long-term cancer survivors' quality of life and self-advocacy. *Communication Research Reports, 32*, 373–378. doi: 10.1080/08824096.2015.1089852

Bowden, C., & Galindo-Gonzalez, S. (2015). Interviewing when you're not face-to-face: The use of email interviews in a phenomenological study. *International Journal of Doctoral Studies, 10*, 79–92. Retrieved from: http://ijds.org/Volume10/IJDSv10p079-092Bowden0684.pdf

Bowker, G. C., & Star, S. L. (2000). *Sorting things out: Classification and its consequences.* Cambridge, MA: MIT Press.

Boylorn, R. M. (2011). Gray or for colored girls who are tired of chasing rainbows: Race and reflexivity. *Cultural Studies ↔ Critical Methodologies, 11*, 178–186. doi: 10.1177/1532708611401336

Brady, I. (2004). In defense of the sensual: Meaning construction in ethnography and poetics. *Qualitative Inquiry, 10*, 622–644. doi: 10.1177/1077800404265719

Brady, J. (2011). Cooking as inquiry: A method to stir up prevailing ways of knowing food, body, and identity. *International Journal of Qualitative Methods, 10*, 321–334. Retrieved from: http://ejournals.library.ualberta.ca/index.php/IJQM/article/view/11580/0

Brown, L., & Boardman, F. K. (2010). Accessing the field: Disability and the research process. *Social Science and Medicine, 72*, 23–30. doi: 10.1016/j.socscimed.2010.09.050

Browne, A. L. (2016). Can people talk together about their practices? Focus groups, humour and the sensitive dynamics of everyday life. *Area, 48*, 198–205. doi: 10.1111/area.12250

Brunyé, T. T., Gagnon, S. A., Waller, D., Hodgson, E., Tower-Richardi, S., & Taylor, H. A. (2012). Up north and down south: Implicit associations between topography and cardinal direction. *The Quarterly Journal of Experimental Psychology, 65*, 1880–1894.

Bryson, M. K., & Stacey, J. (2013). Cancer knowledge in the plural: Queering the biopolitics of narrative and affective mobilities. *Journal of Medical Humanities, 34*, 197–212. doi: 10.1007/s10912-013-9206-z

Bunds, K. S. (2014). The biopolitics of privilege: Negotiating class, masculinity, and relationships. *Cultural Studies ↔ Critical Methodologies, 14*, 517–525. doi: 10.1177/1532708614541895

Burke, K. (1971). *Literature as equipment for living: Critical theory since Plato.* New York: Harcourt.

Burke, K. (1973). *The philosophy of literary form: Studies in symbolic action.* Berkeley, CA: University of California Press.

Burke, L. A., & Miller, M. K. (2001). Phone interviewing as a means of data collection: Lessons learned and practical recommendations. In *Forum Qualitative Sozialforschung/Forum: Qualitative Social Research, 2*(2). Retrieved from: http://www.qualitative-research.net/index.php/fqs/article/viewArticle/959

Burns, M. (2003). Interviewing: Embodied communication. *Feminism & Psychology, 13,* 229–236. doi: 10.1177/0959353503013002006

Butler, J. (1990). Performative acts and gender constitution: An essay in phenomenology and feminist theory. In S. E. Case (Ed.), *Performing feminisms: Feminist critical theory and theatre* (pp. 270–282). Baltimore, MD: Johns Hopkins University Press.

Butler, J. (1997). Performative acts and gender constitution: An essay in phenomenology and feminist theory. In K. Conboy, N. Medina, & S. Stanbury (Eds.), *Writing on the body: Female embodiment and feminist theory* (pp. 401–417). New York: Columbia University Press.

Brison, S. J. (1997). Outliving oneself: Trauma, memory and personal identity. In D. T. Meyers (Ed.), *Feminists rethink the self* (pp. 12–39). Boulder, CO: Westview.

Brooks, C. (2010). Embodied transcription: A creative method for using voice-recognition software. *The Qualitative Report, 15,* 1227–1241. Retrieved from: http://search.proquest.com/docview/757177671?pq-origsite=gscholar

Cahnmann-Taylor, M., & Siegesmund, R. (Eds.). (2013). *Arts-based research in education: Foundations for practice.* New York: Routledge.

Cameron, J. (2002). *The artist's way: A spiritual path to higher creativity.* New York: Tarcher.

Cann, C. N., & DeMeulenaere, E. J. (2012). Critical co-constructed autoethnography. *Cultural Studies ↔ Critical Methodologies, 12,* 146–158. doi: 10.1177/1532708611435214

Carr, J. M. (2003). Poetic expressions of vigilance. *Qualitative Health Research, 13,* 1324–1331. doi: 10.1177/1049732303254018

Carrington, B. (2008). "What's the footballer doing here?" Racialized performativity, reflexivity, and identity. *Cultural Studies ↔ Critical Methodologies, 8,* 423–452. doi: 10.1177/1532708608321574

Charmaz, K. (2006). *Constructing grounded theory: A practical guide through qualitative analysis.* Thousand Oaks, CA: Sage.

Chawla, D. (2008). Poetic arrivals and departures: Bodying the ethnographic field in verse. *Qualitative Sozialforschung/Forum: Qualitative Social Research, 9*(2). Retrieved from http://nbn-resolving.de/urn:nbn:de:0114-fqs0802248

Cherrington, J., & Watson, B. (2010). Shooting a diary, not just a hoop: Using video diaries to explore the embodied everyday contexts of a university basketball team. *Qualitative Research in Sport and Exercise, 2,* 267–281. doi: 10.1080/19398441.2010.488036

Childers, S. M. (2013). The materiality of fieldwork: An ontology of feminist becoming. *International Journal of Qualitative Studies in Education, 26,* 599–609. doi: 10.1080/09518398.2013.786845

Childers, S. M. (2014). Promiscuous analysis in qualitative research. *Qualitative Inquiry, 20,* 819–826. doi: 10.1177/1077800414530266

Cixous, H., & Calle-Gruber, M. (1997). *Hélène Cixous, rootprints: Memory and life writing.* Hove, UK: Psychology Press.

Clair, R. P. (1998). *Organizing silence: A world of possibilities.* Albany, NY: SUNY Press.

Clarke, A. E. (2003). Situational analyses: Grounded theory mapping after the postmodern turn. *Symbolic interaction, 26,* 553–576. doi: 10.1525/si.2003.26.4.553

Clarke, A. E. (2005). *Situational analysis: Grounded theory after the postmodern turn.* Thousand Oaks, CA: Sage.

Clarke, N. J., Willis, M. E., Barnes, J. S., Caddick, N., Cromby, J., McDermott, H., & Wiltshire, G. (2015). Analytical pluralism in qualitative research: A meta-study. *Qualitative Research in Psychology, 12*, 182–201. doi: 10.1080/14780887.2014.948980

Coemans, S., Wang, Q., Leysen, J., & Hannes, K. (2015). The use of arts-based methods in community-based research with vulnerable populations: Protocol for a scoping review. *International Journal of Educational Research, 71*, 33–39. doi: 10.1016/j.ijer.2015. 02.008

Coffey, A. (1999). *The ethnographic self: Fieldwork and the representation of identity*. Thousand Oaks, CA: Sage.

Collins, P. H. (2002). *Black feminist thought: Knowledge, consciousness, and the politics of empowerment* (2nd ed.). New York: Routledge.

Colombetti, G., & Thompson, E. (2008). The feeling body: Toward an enactive approach to emotion. In W. F. Overton, U. Muller, & J. L. Newman (Eds.), *Developmental perspectives on embodiment and consciousness* (pp. 45–67). New York: Erlbaum.

Conboy, K., Medina, N., & Stanbury, S. (Eds.). (1997). *Writing on the body: Female embodiment and feminist theory*. New York: Columbia University Press.

Connell, C. (2011). The politics of the stall: Transgender and genderqueer workers negotiating 'the bathroom question.' In C. Bobel & S. Kwan (Eds.), *Embodied resistance: Breaking rules in public spaces* (pp. 175–185). Nashville, TN: Vanderbilt University Press.

Conrad, D. (2012). *Athabasca's going unmanned: An ethnodrama about incarcerated youth*. Rotterdam, Netherlands: Sense.

Cook, C. (2012). Email interviewing: Generating data with a vulnerable population. *Journal of Advanced Nursing, 68*, 1330–1339. doi: 10.1111/j.1365-2648.2011.05843.x

Corbin, J., & Morse, J. M. (2003). The unstructured interactive interview: Issues of reciprocity and risks when dealing with sensitive topics. *Qualitative Inquiry, 9*, 335–354. doi: 10.1177/1077800403009003001

Covarrubias, P. (2007). (Un)biased in Western theory: Generative silence in American Indian communication. *Communication Monographs, 74*, 265–271. doi: 10.1080/ 03637750701393071

Crawford, C. (2009). From pleasure to pain: The role of the MPQ in the language of phantom limb pain. *Social Science & Medicine, 69*, 655–661. doi: 10.1016/j.socsci med.2009.02.022

Crawford, C. (2014). Body image, prostheses, phantom limbs. *Body & Society, 21*, 1–24. doi: 10.1177/1357034X14522102

Crenshaw, K. (1989). Demarginalizing the intersection of race and sex: A black feminist critique of antidiscrimination doctrine, feminist theory, and antiracist politics. *University of Chicago Legal Forum, 14*, 538–554.

Crenshaw, K. (1991). Mapping the margins: Intersectionality, identity politics, and violence against women of color. *Stanford Law Review, 43*, 1241–1299. Retrieved from: http://www.jstor.org/stable/1229039

Creswell, J. W., & Clark, V. L. P. (2010). *Designing and conducting mixed methods research* (2nd ed.). Thousand Oaks, CA: Sage.

Damianakis, T., & Woodford, M. R. (2012). Qualitative research with small connected communities generating new knowledge while upholding research ethics. *Qualitative Health Research, 22*, 708–718. doi: 10.1177/1049732311431444

Danesi, M. (1999). *Sign, thought, and culture*. Toronto, ON: Canadian Scholars' Press.

Davidson, C. R. (2009). Transcription: Imperatives for qualitative research. *International Journal of Qualitative Methods, 8*(2), 35–52. doi: 10.1177/160940690900800206

Davis, K. (2008). Intersectionality as buzzword: A sociology of science perspective on what makes a feminist theory successful. *Feminist Theory, 9*, 67–85. doi: 10.1177/146470010808636

Day, A. M. (2010). Hide and seek: Exploring connections between embodied rhythmic associations of experiences and of stories. *Qualitative Inquiry, 16*, 697–704. doi: 10.1177/1077800410374181

Daza, S. L., & Huckaby, M. F. (2014). Terra incognita: Em-bodied data analysis. *Qualitative Inquiry, 20*, 801–810. doi: 10.1177/1077800414530264

Defenbaugh, N. L. (2011). *Dirty tale: A narrative journey of the IBD body.* Cresskill, NJ: Hampton.

Defenbaugh, N. L. (2013). Revealing and concealing ill identity: A performance narrative of IBD disclosure. *Health Communication, 28*, 159–169. doi: 10.1080/10410236.2012.666712

de Freitas, E. (2012). The classroom as rhizome: New strategies for diagramming knotted interactions. *Qualitative Inquiry, 18*, 557–570. doi: 10.1177/1077800412450155

Del Busso, L. (2007). Embodying feminist politics in the research interview: Material bodies and reflexivity. *Feminism Psychology, 17*, 309–315. doi: 10.1177/0959353507079084

Deleuze, G. (1993). *The fold: Leibniz and the baroque* (T. Conley, Trans.). Minneapolis, MN: University of Minnesota Press.

Deleuze, G. (1994). *Difference and repetition* (P. Patton, Trans.). New York: Columbia University Press.

Deleuze, G., & Guattari, F. (1987). *A thousand plateaus: Capitalism and schizophrenia.* (B. Massumi, Trans.). Minneapolis, MN: University of Minnesota Press.

Deleuze, G., & Guattari, F. (1994). What is philosophy? (H. Tomlinson & G. Burchell, Trans.). New York: Columbia University Press.

Denham, M. A., & Onwuegbuzie, A. J. (2013). Beyond words: Using nonverbal communication data in research to enhance thick description and interpretation. *International Journal of Qualitative Methods, 12*, 670–696. Retrieved from: https://ejournals.library.ualberta.ca/index.php/IJQM/article/view/19271

Denshire, S. (2014). On auto-ethnography. *Current Sociology, 62*, 831–850. doi: 10.1177/0011392114533339

Denshire, S., & Lee, A. (2013). Conceptualizing autoethnography as assemblage: Accounts of occupational therapy practice. *International Journal of Qualitative Methods, 12*, 221–236. doi: 10.1177/160940691301200110

Denzin, N. K. (1997). *Interpretive ethnography: Ethnographic practices for the 21st century.* Thousand Oaks, CA: Sage.

Denzin, N. K., & Lincoln, Y. (2011). *Sage handbook of qualitative research* (4th ed.). Thousand Oaks, CA: Sage.

DeVault, M. L. (1990). Talking and listening from women's standpoint: Feminist strategies for interviewing and analysis. *Social problems, 37*, 96–116. doi: http://dx.doi.org/10.2307/800797

De Zordo, S. (2012). Programming the body, planning reproduction, governing life: The '(ir-) rationality' of family planning and the embodiment of social inequalities in Salvador da Bahia (Brazil). *Anthropology & Medicine, 19*, 207–223. doi: 10.1080/13648470.2012.675049

Dillaway, H. E. (2011). Menopausal and misbehaving: When women "flash" in front of others. In C. Bobel & S. Kwan (Eds.), *Embodied resistance: Challenging the norms, breaking the rules,* (pp. 197–208). Nashville, TN: Vanderbilt University Press.

Diver, S. W., & Higgins, M. N. (2014). Giving back through collaborative research: Towards a practice of dynamic reciprocity. *Journal of Research Practice, 10*(2), Article M9. Retrieved from: http://jrp.icaap.org/index.php/jrp/article/view/415/354

Dosekun, S. (2015). "Hey, you stylized woman there": An uncomfortable reflexive account of performative practices in the field. *Qualitative Inquiry, 21*, 436–444. doi: 10.1177/1077800415569788

Doucet, A. (2009). Dad and baby in the first year: Gendered responsibilities and embodiment. *The ANNALS of the American Academy of Political and Social Science, 624*, 78–98. doi: 10.1177/0002716209334069

Drabble, L., Trocki, K. F., Salcedo, B., Walker, P. C., & Korcha, R. A. (2016). Conducting qualitative interviews by telephone: Lessons learned from a study of alcohol use among sexual minority and heterosexual women. *Qualitative Social Work, 15*, 118–133. doi: 10.1177/1473325015585613

Draper, J. (2003). Blurring, moving and broken boundaries: Men's encounters with the pregnant body. *Sociology of Health & Illness, 25*, 743–767. doi: 10.1046/j.1467-9566.2003.00368.x

Drew, S. K., Mills, M., & Gassaway, B. M. (2007). *Dirty work: The social construction of taint.* Waco, TX: Baylor University Press.

Dubinskas, F. A. (1988). Cultural constructions: The many faces of time. In F. A. Dubinskas (Ed.), *Making time: Ethnographies of high-technology organizations* (pp. 3–38). Philadelphia, PA: Temple University Press.

Duff, C. (2012). After methods, after subjects, after drugs. *Contemporary Drug Problems, 39*, 265–287. Retrieved from: http://heinonline.org/HOL/Page?handle=hein.journals/condp39&div=15&g_sent=1&collection=journals#

Duranti, A. (2006). Transcripts, like shadows on a wall. *Mind, Culture, and Activity, 13*, 301–310. doi: 10.1207/s15327884mca1304_3

Edley, P. P. (2000). Discursive essentializing in a woman-owned business: Gendered stereotypes and strategic subordination. *Management Communication Quarterly, 14*, 271–306.

Edwards, J. A. (2003). The transcription of discourse. In D. Schiffrin, D. Tannen, & H. E. Hamilton (Eds.), *The handbook of discourse analysis* (pp. 321–348). Malden, MA: Blackwell.

Eisenberg, E. M., & Goodall, H. L. (2004). *Organizational communication: Balancing creativity and constraint..*

Ekman, P., Friesen, W. V., & Ellsworth, P. (2013). *Emotion in the human face: Guidelines for research and an integration of findings.* New York: Pergamon.

Ellingson, L. L. (1998). "Then you know how I feel": Empathy, identification, and reflexivity in fieldwork. *Qualitative Inquiry, 4*, 492–514. doi: 10.1177/107780049800400405

Ellingson, L. L. (2005). *Communicating in the clinic: Negotiating frontstage and backstage teamwork.* Cresskill, NJ: Hampton Press.

Ellingson, L. L. (2006). Embodied knowledge: Writing researchers' bodies into qualitative health research. *Qualitative Health Research, 16*, 298–310. doi: 10.1177/1049732305281944

Ellingson, L. L. (2008). Changing realities and entrenched norms in dialysis: A case study of power, knowledge, and communication in health care delivery. In M. Dutta & H. Zoller (Eds.), *Emerging perspectives in health communication: Meaning, culture, and power* (pp. 293–312). New York: Routledge.

Ellingson, L. L. (2009a). *Engaging crystallization in qualitative research: An introduction.* Thousand Oaks, CA: Sage.

Ellingson, L. L. (2009b). Salvaging, surrendering, and saying goodbye to my leg. *Health Communication, 24*, 773–774. doi: 10.1080/10410230903318497

Ellingson, L. L. (2011a). Analysis and representation across the continuum. In N. K. Denzin & Y. Lincoln, *Sage handbook of qualitative research* (4th ed., pp. 595–610). Thousand Oaks, CA: Sage.

Ellingson, L. L. (2011b). The poetics of professionalism among dialysis technicians. *Health Communication, 26,* 1–12. doi: 10.1080/10410236.2011.527617

Ellingson, L. L. (2012). Interviewing as embodied communication. In J. Gubrium, J. Holstein, A. Marvasti, & K. M. Marvasti (Eds.), *Handbook of interview research* (2nd ed., pp. 525–539). Thousand Oaks, CA: Sage.

Ellingson, L. L. (2015). Embodied practices of dialysis care: On (para)professional work. In B. Green & N. Hopwood (Eds.), *The body in professional practice, learning and education: Body/practice* (pp. 173–189). Cham, Switzerland: Springer.

Ellingson, L. L. (2017). Realistically ever after: Disrupting dominant narratives promoted by cancer advocacy organizations. *Management Communication Quarterly.* doi: 10.1177/0893318917689894

Ellingson, L. L., & Buzzanell, P. M. (1999). Listening to women's narratives of breast cancer treatment: A feminist approach to patient satisfaction with physician-patient communication. *Health Communication, 11,* 153–183. doi: 10.1207/s15327027hc1102_3

Ellingson, L. L., & Ellis, C. (2008). Autoethnography as constructionist project. In J. A. Holstein & J. F. Gubrium (Eds.), *Handbook of constructionist research* (pp. 445–465). New York: Guilford.

Ellingson, L. L., & Quinlan, M. M. (2012). Beyond the research/service dichotomy: Claiming *all* research products for hiring, evaluation, tenure, and promotion. *Qualitative Communication Research, 1,* 385–399. doi: 10.1525/qcr.2012.1.3.385

Ellingson, L. L., & Sotirin, P. (2010). *Aunting: Cultural practices that sustain family and community life.* Waco, TX: Baylor University Press.

Ellis, C. (1991). Sociological introspection and emotional experience. *Symbolic Interaction, 14,* 23–50. doi: 10.1525/si.1991.14.1.23

Ellis, C. (2004). *The ethnographic I: A methodological novel about autoethnography.* Walnut Creek, CA: AltaMira Press.

Ellis, C. (2017). Compassionate research: Interviewing and storytelling from a relational ethics of care. In I. Goodson, M. Andrews, & A. Antikainen (Eds.), *The Routledge international handbook on narrative and life history* (pp. 431–445). New York: Routledge.

Ellis, C., & Bochner, A. P. (2000). Autoethnography, personal narrative, reflexivity: Researcher as subject. In N. K. Denzin & Y. S. Lincoln (Eds.), *Handbook of qualitative research* (2nd ed., pp. 733–768). Thousand Oaks, CA: Sage.

Ellis, C., & Ellingson, L. L. (2000). Qualitative methods. In E. F. Borgatta & R. J. V. Montgomery (Eds.), *Encyclopedia of Sociology* (2nd ed., Vol. 4, pp. 2287–2296). New York: Macmillan.

Ellis, C., & Patti, C. (2014). With heart: Compassionate interviewing and storytelling with holocaust survivors. *Storytelling, Self, Society, 10,* 93–118. doi: 10.13110/storselfsoci.10.1.0093

Ellis, C., & Rawicki, J. (2013). Collaborative witnessing of survival during the Holocaust: An exemplar of relational autoethnography. *Qualitative Inquiry, 19,* 366–380. doi: 10.1177/1077800413479562

El Refaie, E. (2014). Appearances and dis/dys-appearances: A dynamic view of embodiment in Conceptual Metaphor Theory. *Metaphor and the Social World, 4,* 109–125. doi: 10.1075/msw.4.1.08ref

Emerson, R. M., Fretz, R. I., & Shaw, L. L. (2011). *Writing ethnographic fieldnotes.* Chicago, IL: University of Chicago Press.

Enriquez, J. G. (2012). Bodily aware in cyber-research. In H. Breslow & A. Mousoutzanis (Eds.), *Cybercultures: Mediations of community, culture, politics* (pp. 59–72). Amsterdam: Rodopi.

Esposito, J. (2014). Pain is a social construction until it hurts: Living theory on my body. *Qualitative Inquiry, 20*, 1179–1190. doi: 10.1177/1077800414545234

Ezzy, D. (2010). Qualitative interviewing as an embodied emotional performance. *Qualitative Inquiry, 16*, 163–170. doi: 10.1177/1077800409351970

Farman, J. (2015). Stories, spaces, and bodies: The production of embodied space through mobile media storytelling. *Communication Research and Practice, 1*, 101–116. doi: 10.1080/22041451.2015.1047941

Farnsworth, J., & Austrin, T. (2010). The ethnography of new media worlds? Following the case of global poker. *New Media & Society, 12*, 1120–1136. doi: 10.1177/1461444809355648

Faulkner, S. L. (2009). *Poetry as method: Reporting research through verse.* Walnut Creek, CA: Left Coast Press.

Faulkner, S. L. (2014). *Family stories, poetry, and women's work: Knit four, frog one.* Rotterdam: Sense Publishers.

Faulkner, S. L. (2016). TEN (the promise of arts-based, ethnographic, and narrative research in critical family communication research and praxis). *Journal of Family Communication, 16*, 9–15. doi: 10.1080/15267431.2015.1111218

Fenge, L. A., & Jones, K. (2012). Gay and pleasant land? Exploring sexuality, ageing and rurality in a multi-method, performative project. *British Journal of Social Work, 42*, 300–317.

Field, T. M. (2014). *Touch in early development.* Hove, UK: Psychology Press.

Fielding, N. G., Lee, R. M., & Blank, G. (Eds.). (2008). *The SAGE handbook of online research methods.* Thousand Oaks, CA: Sage.

Fine, G. A. (1993). Ten lies of ethnography: Moral dilemmas of field research. *Journal of Contemporary Ethnography, 22*, 267–294. doi: 10.1177/089124193022003001

Fine, G. A., & Deegan, J. G. (1996). Three principles of serendip: Insight, chance, and discovery in qualitative research. *International Journal of Qualitative Studies in Education, 9*, 434–447. doi: 10.1080/0951839960090405

Fine, M. (1994). Working the hyphens: Reinventing self and other in qualitative research. In N. K. Denzin & Y. S. Lincoln (Eds.), *Handbook of qualitative research* (pp. 70–82). Thousand Oaks, CA: Sage.

Fine, M., & McClelland, S. (2006). Sexuality education and desire: Still missing after all these years. *Harvard Educational Review, 76*, 297–338. doi: http://dx.doi.org/10.17763/haer.76.3.w5042g23122n6703

Finley, N. J. (2010). Skating femininity: Gender maneuvering in women's roller derby. *Journal of Contemporary Ethnography, 39*, 359–387. doi: 10.1177/0891241610364230

Finley, S. (2015). Embodied homelessness: The pros/thesis of art research. *Qualitative Inquiry, 21*, 504–509. doi: 10.1177/1077800415581886

Flanagan, M. K. (2014). Sporting a skort: The biopolitics of materiality. *Cultural Studies ↔ Critical Methodologies, 14*, 506–516. doi: 10.1177/1532708614541894

Foucault, M. (1977). *Discipline and punish: The birth of the prison.* New York: Vintage.

Fox, K. M., Humberstone, B., & Dubnewick, M. (2014). Cycling into sensoria: Embodiment, leisure, and tourism. *Tourism Review International, 18*, 71–85. doi: 10.3727/154427214X13990420684563

Fox, K. V. (1996). Silent voices: A subversive reading of child sexual abuse. In C. Ellis & A. P. Bochner (Eds.), *Composing ethnography: Alternative forms of qualitative writing* (pp. 330–356). Walnut Creek, CA: AltaMira.

Francombe-Webb, J., & Silk, M. (2016). Young girls' embodied experiences of femininity and social class. *Sociology, 50*, 652–672. doi: 0.1177/0038038514568233

Frank, A. W. (1995). *The wounded storyteller: Body, illness, and ethics*. Chicago, IL: University of Chicago Press.

Frank, A. W. (2000). Illness and autobiographical work: Dialogue as narrative destabilization. *Qualitative Sociology, 23*, 135–156. doi: 10.1023/A:1005411818318

Frith, H. (2015). Visualising the 'real' and the 'fake': Emotion work and the representation of orgasm in pornography and everyday sexual interactions. *Journal of Gender Studies, 24*, 386–398. doi: 10.1080/09589236.2014.950556

Furman, R. (2006). Poetic forms and structures in qualitative health research. *Qualitative Health Research, 16*, 560–566. doi: 10.1177/1049732306286819

Gale, N. K. (2010). The embodied ethnographer: Journeys in a health care subculture. *International Journal of Qualitative Methods, 9*, 206–223. doi: 10.1177/160940691000900206

Gallagher, M. (2011). Sound, space and power in a primary school. *Social & Cultural Geography, 12*, 47–61. doi: 10.1080/14649365.2011.542481

Garratt, D. (2015). Queer and uncanny: An ethnographic critique of female natural bodybuilding. *Qualitative Inquiry, 21*, 776–786. doi: 10.1177/1077800415574910

Geertz, C. (1973). *The interpretation of cultures*. New York: Basic Books.

Gemignani, M. (2014). Memory, remembering, and oblivion in active narrative interviewing. *Qualitative Inquiry, 20*, 127–135. doi: 10.1177/1077800413510271

Gergen, K. J. (1994). *Realities and relationships: Soundings in social construction*. Cambridge, MA: Harvard University Press.

Gibbs, L. (2014). Arts-science collaboration, embodied research methods, and the politics of belonging: 'SiteWorks' and the Shoalhaven River, Australia. *Cultural Geographies, 21*, 207–227. doi: 10.1177/1474474013487484

Gibson, J. J. (1966). *The senses considered as perceptual systems*. Boston, MA: Houghton Mifflin.

Gibson, J. J. (1979). *The ecological approach to visual perception*. Boston, MA: Houghton Mifflin.

Gilbert, E., Ussher, J. M., & Perz, J. (2013). Embodying sexual subjectivity after cancer: A qualitative study of people with cancer and intimate partners. *Psychology & Health, 28*, 603–619. doi: 10.1080/08870446.2012.737466

Gingrich-Philbrook, C. (2005). Autoethnography's family values: Easy access to compulsory experiences. *Text and Performance Quarterly, 25*, 297–314. doi: 10.1080/10462930500362445

Glass, N., & Ogle, K. R. (2012). Embodiment of the interpersonal nexus: Revealing qualitative research findings on shoulder surgery patients. *Journal of Multidisciplinary Healthcare, 5*, 69–76. doi: 10.2147/JMDH.S29273

Glennie, E. (1993). Hearing essay. Retrieved from: https://www.evelyn.co.uk/hearing-essay/

Glesne, C. (1997). That rare feeling: Re-presenting research through poetic transcription. *Qualitative Inquiry, 3*, 202–221. doi: 10.1177/107780049700300204

Goffman, E. (1959). *The presentation of self in everyday life*. Garden City, NY: Doubleday.

Goffman, E. (1963). *Behavior in public places: Notes on the social organization of gatherings*. New York: Free Press.

Goldberg, N. (1986). *Writing down the bones: Freeing the writer within*. Boston, MA: Shambhala.

Goodley, D., & Runswick-Cole, K. (2013). Disability: Cripping men, masculinities and methodologies. In B. Pini & B. Pease (Eds.), *Men, masculinities and methodologies* (pp. 142–156). London, UK: Palgrave Macmillan.

Grauerholz, L., Barringer, M., Colyer, T., Guittar, N., Hecht, J., Rayburn, R. L., & Swart, E. (2013). Attraction in the field: What we need to acknowledge and implications for research and teaching. *Qualitative Inquiry, 19*, 167–178. doi: 10.1177/1077800412466222

Gray, R., & Sinding, C. (2002). *Standing ovation: Performing social science research about cancer.* Walnut Creek, CA: AltaMira Press.

Green, J., Franquiz, M., & Dixon, C. (1997). The myth of the objective transcript: Transcribing as a situated act. *Tesol Quarterly, 31*, 172–176. doi: 10.2307/3587984

Greiner, K. (2012). Participatory health communication research. In R. Obregon & S. Waisbord (Eds.), *The handbook of global health communication* (pp. 348–373). Hoboken, NJ: Wiley-Blackwell.

Grosz, E. A. (1994). *Volatile bodies: Toward a corporeal feminism.* Bloomington, IN: Indiana University Press.

Guba, E. G., & Lincoln, Y. S. (1994). Competing paradigms in qualitative research. In N. K. Denzin & Y. Lincoln (Eds.), *Handbook of qualitative research* (pp. 105–117). Thousand Oaks, CA: Sage.

Gubrium, J. F., & Holstein, J. A. (1997). *The new language of qualitative method.* New York: Oxford University Press.

Gunaratnam, Y. (2003). *Researching race and ethnicity: Methods, knowledge and power.* London, UK: Sage.

Guntram, L. (2013). "Differently normal" and "normally different": Negotiations of female embodiment in women's accounts of 'atypical' sex development. *Social Science & Medicine, 98*, 232–238. doi: 10.1016/j.socscimed.2013.09.018

Guzzardo, M. T., Adams, W. E., Todorova, I. L., & Falcón, L. M. (2016). Resonating sentiments on Puerto Rican identity through poetry voices of the diaspora. *Qualitative Inquiry, 22*, 428–443. doi: 10.1177/1077800415622485

Hall, A., Hockey, J., & Robinson, V. (2007). Occupational cultures and the embodiment of masculinity: Hairdressing, estate agency and firefighting. *Gender, Work & Organization, 14*(6), 534–551. doi: 10.1111/j.1468-0432.2007.00370.x

Hall, T., Lashua, B., & Coffey, A. (2008). Sound and the everyday in qualitative research. *Qualitative Inquiry, 14*, 1019–1040. doi: 10.1177/1077800407312054

Haraway, D. (1988). Situated knowledges: The science question in feminism and the privilege of partial perspective. *Feminist Studies, 14*, 575–599. doi: 10.2307/3178066

Haraway, D. (1991). *Simians, cyborgs and women: The reinvention of nature.* New York: Routledge, and London, UK: Free Association Books.

Harding, S. G. (1991). *Whose science? Whose knowledge?: Thinking from women's lives.* Ithaca, NY: Cornell University Press.

Harley, J. B. (1989). Deconstructing the map. *Cartographica: The International Journal for Geographic Information and Geovisualization, 26*(2), 1–20.

Harris, M. (2009). Injecting, infection, illness: Abjection and hepatitis C stigma. *Body & Society, 15*(4), 33–51. doi: 10.1177/1357034X09347221

Harris, M. (2012). *Harm reduction and me: Exchange supplies.* Retrieved from http://www.youtube.com/ watch?v=0H51Nk-I7PA

Harris, M. (2015). "Three in the room": Embodiment, disclosure, and vulnerability in qualitative research. *Qualitative Health Research, 25*, 1689–1699. doi: 10.1177/1049732314566324

Harris, A., & Guillemin, M. (2012). Developing sensory awareness in qualitative interviewing: A portal into the otherwise unexplored. *Qualitative Health Research, 22*, 689–699. doi: 10.1177/1049732311431899

Harter, L. M. (2013a). The work of art. *Departures in Critical Qualitative Research, 2,* 326–336. doi: 10.1525/qcr.2013.2.3.326

Harter, L. M. (2013b). *Imagining new normals: A narrative framework for health communication.* Dubuque, IA: Kendall Hunt.

Harter, L. M., & Hayward, C. (Producers). (2010). *The art of the possible* [film]. Athens, OH: Ohio University Scripps College of Communication.

Harter, L. M., Patterson, S., & Gerbensky-Kerber, A. (2010). Narrating "new normals" in health care contexts. *Management Communication Quarterly, 24,* 465–473. doi: 10.1177/0893318910370271

Harter, L., Hamel-Lambert, J., & Millesen, J. (Eds.). (2011). *Participatory partnerships for social action and research.* Dubuque, IA: Kendall Hunt.

Harter, L. M., Shaw, E. (Producers), Quinlan, M. M., Ruhl, S. M., & Hodson, T. (Assistant Producers). (2015). *A beautiful remedy* [film]. Athens, OH: WOUB Public Media.

Harter, L. M., Shaw, E., & Quinlan, M. M. (Producers). (2016). *Creative abundance* [film]. Athens, OH: WOUB Center for Public Media.

Hartnett, S. J. (2003). *Incarceration nation: Investigative prison poems of hope and terror.* Walnut Creek, CA: AltaMira Press.

Hawkins, J. M. (in press). Transcription systems. In M. Allen (Ed.), *Encyclopedia of communication research methods.* Thousand Oaks, CA: Sage.

Hekman, S. (2010). *The material of knowledge: Feminist disclosures.* Bloomington, IN: Indiana University Press.

Hesse-Biber, S. N. (Ed.). (2011). *Handbook of feminist research: Theory and praxis* (2nd ed.). Thousand Oaks, CA: Sage.

Hesse-Biber, S. N., & Brooks, A. (2007). Core feminist insights and strategies on authority, representation, truths, reflexivity, and ethics across the research process. In S. N. Hesse-Biber (Ed.), *Handbook of feminist research: Theory and praxis* (pp. 419–424). Thousand Oaks, CA: Sage

Hickson, M. L., Stacks, D. W., & Moore, N. J. (2004). *Nonverbal communication: Studies and applications.* Los Angeles, CA: Roxbury Publishing.

Hiebert, V., & Hiebert, D. W. (2015). Intersex persons and the church: Unknown, unwelcomed, unwanted neighbors. *Journal for the Sociological Integration of Religion and Society, 5*(2), 31–44. Retrieved from: http://www.religionandsociety.org/ojs/index.php/jsirs/article/view/111

Hightower, J. L. (2015). Producing desirable bodies: Boundary work in a lesbian niche dating site. *Sexualities, 18,* 20–36. doi: 10.1177/1363460714550900

Hochschild, A. R. (1983). *The managed heart: Commercialization of human feeling.* Berkeley, CA: University of California Press.

Hockey, J., & James, A. (2012). Health and the embodiment of the life course. In B. S. Turner (Ed.), *Routledge handbook of body studies* (pp. 275–285). New York: Routledge.

Hoel, N. (2013). Embodying the field: A researcher's reflections on power dynamics, positionality, and the nature of research relationships. *Fieldwork in Religion, 8,* 27–49. doi: 10.1558/fiel.v8i1.27

Holman Jones, S. (2016). Living bodies of thought: The "critical" in critical autoethnography. *Qualitative Inquiry, 22,* 228–237. doi: 10.1177/1077800415622509

Holman Jones, S., Adams, T. E., & Ellis, C. (Eds.). (2016). *Handbook of autoethnography.* New York: Routledge.

Holmes, R. (2014). Fresh kills: The spectacle of (de)composing data. *Qualitative Inquiry, 20,* 781–789. doi: 10.1177/1077800414530262

Honan, E. (2014). Disrupting the habit of interviewing. *Reconceptualizing Educational Research Methodology, 5*(1). doi: 10.7577/rerm.929. Retrieved from: http://espace.library.uq.edu.au/view/UQ:338699/disrupting_interviewing.pdf

Honkasalo, M. (2001). Vicissitudes of pain and suffering: Chronic pain and liminality. *Medical Anthropology, 19*, 319–535. doi: 10.1080/01459740.2001.9966181

hooks, b. (2000). *Feminist theory: From margin to center*. London, UK: Pluto Press.

Hopkins, J. B. (2015). Coming "home": An autoethnographic exploration of third culture kid transition. *Qualitative Inquiry, 21*, 812–820. doi: 10.1177/1077800415574909

Hopwood, N. (2013). Ethnographic fieldwork as embodied material practice: Reflections from theory and the field. *Studies in Symbolic Interaction, 40*, 227–245. doi: 10.1108/S0163-2396(2013)0000040013

Horner, J. (2010). Targeted bodies. *International Journal of Communication, 4*, 241–245. Retrieved from: http://ijoc.org/index.php/ijoc/article/view/738/0

Horowitz, C. R., Robinson, M., & Seifer, S. (2009). Community-based participatory research from the margin to the mainstream: Are researchers prepared? *Circulation, 119*, 2633–2642. doi: 10.1161/CIRCULATIONAHA.107.729.863

Hudak, P. L., McKeever, P., & Wright, J. G. (2007). Unstable embodiments: A phenomenological interpretation of patient satisfaction with treatment outcome. *Journal of Medical Humanities, 28*, 31–44. doi: 10.1007/s10912-006-9027-4

Ihde, D. (2002). *Bodies in technology*. Minneapolis, MN: University of Minnesota Press.

Inckle, K. (2014). A lame argument: Profoundly disabled embodiment as critical gender politics. *Disability & Society, 29*, 388–401. doi: 10.1080/02674649266780451

Ingold, T. (2000). *The perception of the environment: Essays on livelihood, dwelling and skill*. Hove, UK: Psychology Press.

Irigaray, L. (2001). *To be two* (M. Rhodes & M. Cocito-Monoc, Trans.). New York: Routledge. (Original work published 1994).

Jackson, A. Y., & Mazzei, L. A. (2013). Plugging one text into another: Thinking with theory in qualitative research. *Qualitative Inquiry, 19*, 261–271. doi: 10.1177/1077800412471510

Jackson, S., & Scott, S. (2010). *Theorizing society*. Milton Keynes, UK: Open University Press.

Jago, B. J. (2002). Chronicling an academic depression. *Journal of Contemporary Ethnography, 31*, 729–757. doi: 10.1177/089124102237823

James, N. (2007). The use of email interviewing as a qualitative method of inquiry in educational research. *British Educational Research Journal, 33*, 963–976. doi: 10.1080/01411920701657074

James, N., & Busher, H. (2012). Internet interviewing. In J. Gubrium (Ed.), *The SAGE handbook of interview research: The complexity of the craft* (pp. 177–188). Thousand Oaks, CA: Sage.

Janesick, V. J. (2000). The choreography of qualitative research design: Minuets, improvisations, and crystallization. In N. K. Denzin & Y. S. Lincoln (Eds.), *Handbook of qualitative research* (2nd ed., pp. 379–399). Thousand Oaks, CA: Sage.

Jay, M. (1994). *Downcast eyes: The denigration of vision in twentieth century French thought*. Berkeley, CA: University of California Press.

Johnson, B. & Quinlan, M. M. (2016). Insiders and outsiders and insider(s) again in the (in)fertility world. *Health Communication*. Online First. doi: 10.1080/10410236.2016.1138384.

Johnson, C. S., & Eaves, K. L. (2013). An ounce of time, a pound of responsibilities and a ton of weight to lose: An autoethnographic journey of barriers, message adherence and the weight-loss process. *Public Relations Inquiry, 21*, 95–116. doi: 10.1177/2046147X12460949

Johnson, M. (1987). *The body in the mind: The bodily basis of meaning, imagination, and cognition.* Chicago, IL: University of Chicago Press.

Jones, A., & Woolley, J. (2015). The email-diary: A promising research tool for the 21st century? *Qualitative Research, 15,* 705–721. doi: 10.1177/1468794114561347

Jones, S., & Woglom, J. F. (2015). Behind the body-filled scenes: Methodologies at work on the body in graphica. In M. Perry & C. L. Medina (Eds.), *Methodologies of embodiment: Inscribing bodies in qualitative research* (pp. 116–137). New York: Routledge.

Jones, C. A., Arning, B., & Farver, J. (2006). *Sensorium: Embodied experience, technology, and contemporary art.* Cambridge, MA: MIT Press.

Kay, S. (2011). If I should have a daughter . . . TED Talk. Retrieved from: https://www.ted.com/talks/sarah_kay_if_i_should_have_a_daughter?language=en

Keeler, S. (2012). First do no harm? Female hysteria, trauma, and the (bio)logic of violence in Iraq. *Medical Anthropology, 31,* 132–148. doi: 10.1080/01459740.2011.622152

Keilty, P. (2016). Embodied engagements with online pornography. *The Information Society, 32,* 64–73. doi: 10.1080/01972243.2015.1107162

Kelly, D. (2009). Changed men: The embodied impact of prostate cancer. *Qualitative Health Research, 19,* 151–163. doi: 10.1177/1049732308328067

Kennedy, B. L. (2009). Infusing participants' voices into grounded theory research: A poetic anthology. *Qualitative Inquiry, 15,* 1416–1433. doi: 10.1177/1077800409339569

Kenny, C. (2011). *The power of silence: Silent communication in daily life.* London, UK: Karnac Books.

Kimmel, M. S. (2004). Masculinity as homophobia: Fear, shame, and silence in the construction of gender identity. In P. S. Rothenberg (Ed.), *Race, class, and gender in the United States: An integrated study* (6th ed., pp. 81–93). New York: Worth.

King, T. J. (2007). Bad habits and prosthetic performances: Negotiation of individuality and embodiment of social status in Australian shark fishing. *Journal of Anthropological Research, 63,* 537–560. Retrieved from: http://www.jstor.org/stable/20479464

Koelsch, L. E. (2013). Reconceptualizing the member check interview. *International Journal of Qualitative Methods, 12,* 168–179. doi: 10.1177/160940691301200105

Kondo, D. K. (1990). *Crafting selves: Power, gender, and discourses of identity in a Japanese workplace.* Chicago, IL: University of Chicago Press.

Konopásek, Z. (2008). Making thinking visible with atlas.ti: Computer assisted qualitative analysis as textual practices. *Forum: Qualitative Social Research, 9*(2), 1–21. Retrieved from: http://www.qualitative-research.net/index.php/fqs/article/view/420/910

Kontos, P. C. (2011). Rethinking sociability in long-term care: An embodied dimension of selfhood. *Dementia, 11,* 329–346. doi: 10.1177/1471301211421073

Kontos, P. C., & Naglie, G. (2007). Expressions of personhood in Alzheimer's disease: An evaluation of research-based theatre as a pedagogical tool. *Qualitative Health Research, 17,* 799–811. doi: 10.1177/1049732307302838

Koro-Ljungberg, M., & MacLure, M. (2013). Provocations, re-un-visions, death, and other possibilities of "data". *Cultural Studies ↔ Critical Methodologies, 13,* 219–222. doi: 10.1177/1532708613487861

Korzybski, A. (1933). A non-Aristotelian system and its necessity for rigour in mathematics and physics. In A. Korzybski, *Science and sanity: An introduction to non-Aristotelian systems and general semantics* (pp. 747–762). Lakeville, CT: International Non-Aristotelian Library.

Kovach, M. (2015). Emerging from the margins: Indigenous methodologies. In S. Strega & L. Brown (Eds.), *Research as resistance: Revisiting critical, indigenous, and anti-oppressive approaches* (2nd. ed., pp. 43–64). Toronto, ON: Canadian Scholars' Press.

Kuntz, A. M., & Presnall, M. M. (2012). Wandering the tactical: From interview to intraview. *Qualitative Inquiry, 18*, 732–744. doi: 1077800412453016.

Kusenbach, M. (2003). Street phenomenology: The go-along as ethnographic research tool. *Ethnography, 4*, 455–485. doi: 10.1177/146613810343007

Lahman, M. K., & Richard, V. M. (2014). Appropriated poetry: Archival poetry in research. *Qualitative Inquiry, 20*, 344–355. doi: 10.1177/1077800413489272

La Jevic, L., & Springgay, S. (2008). A/r/tography as an ethics of embodiment: Visual journals in preservice education. *Qualitative Inquiry, 14*, 67–89. doi: 10.1177/1077800407304509

Lasén, A., & García, A. (2015). ' . . . but I haven't got a body to show': Self-pornification and male mixed feelings in digitally mediated seduction practices. *Sexualities, 18*, 714–730. doi: 10.1177/1363460714561720

Lather, P. (1986). Issues of validity in openly ideological research: Between a rock and a soft place. *Interchange, 17*(4), 63–84. doi: 10.1007/bf01807017

Lather, P. (1991). *Getting smart: Feminist research and pedagogy with/in the postmodern.* New York: Routledge.

Lather, P. (1993). Fertile obsession: Validity after poststructuralism. *Sociological Quarterly, 34*, 673–693. doi: 10.1111/j.1533-8525.1993.tb00112.x

Lather, P., & Smithies, C. (1997). *Troubling the angels: Women living with HIV/AIDS.* Boulder, CO: Westview Press.

Latour, B. (2003). A strong distinction between humans and non-humans is no longer required for research purposes: A debate between Bruno Latour and Steve Fuller. *History of the Human Sciences, 16*, 77–99. doi: 10.1177/0952695103016002004

Latour, B. (2005). *Reassembling the social: An introduction to actor-network theory.* Oxford, UK: Oxford University Press.

Law, J. (2004). *After method: Mess in social science research.* London, UK: Routledge.

Law, J. (2007). Making a mess with method. In W. Outhwaite & S. P. Turner (Eds.), *The Sage handbook of social science methodology* (pp. 595–606). London, UK: Sage.

Leavy, P. (2015). *Method meets art: Arts-based research practice.* New York: Guilford.

Leder, D. (1990). *The absent body.* Chicago, IL: University of Chicago Press.

Ledger, A., & McCaffrey, T. (2015). Performative, arts-based, or arts-informed? Reflections on the development of arts-based research in music therapy. *Journal of Music Therapy, 52*, 441–456. doi: 10.1093/jmt/thv013

LeFrançois, B. A., Menzies, R., & Reaume, G. (Eds.). (2013). *Mad matters: A critical reader in Canadian mad studies.* Toronto, ON: Canadian Scholars' Press.

LeGreco, M. (2012). Working with policy: Restructuring healthy eating practices and the Circuit of Policy Communication. *Journal of Applied Communication Research, 40*, 44–64. doi: 10.1080/00909882.2011.636372

LeGreco, M. & Leonard, D. (2011). Building sustainable community-based food programs: Cautionary tales from "The Garden". *Environmental Communication, 5*, 356–362. doi: 10.1080/17524032.2011.593639

LeGreco, M., Leonard, D., & Ferrier, M. (2012). Virtual vines: Using participatory methods to connect virtual work with community-based practice. In S. D. Long (Ed.), *Virtual work and human interaction* (pp. 78–98). Hershey, PA: IGI Global.

Leininger M. (2002). The theory of culture care and the ethnonursing research method. In M. Leininger & M. R. McFarland (Eds.), *Transcultural nursing: Concepts, theories, research and practice* (3rd ed., pp. 71–98). New York: McGraw-Hill.

Liamputtong, P. (2007). *Researching the vulnerable: A guide to sensitive research methods.* Thousand Oaks, CA: Sage.

Lindemann, K. (2012). Access-ability and disability: Performing stigma, writing trauma. *The Northwest Journal of Communication, 40,* 129–149.

Lindlof, T. R., & Taylor, B. C. (2010). *Qualitative communication research methods* (2nd ed.). Thousand Oaks, CA: Sage.

Literat, I. (2013). "A pencil for your thoughts": Participatory drawing as a visual research method with children and youth. *International Journal of Qualitative Methods, 12*(1), 84–98. doi: 10.1177/160940691301200143

Longhurst, R., Ho, E., & Johnston, L. (2008). Using 'the body' as an 'instrument of research': Kimch'i and pavlova. *Area, 40,* 208–217. doi: 10.1111/j.1475-4762.2008.00805.x

López, E. D., Eng, E., Randall-David, E., & Robinson, N. (2005). Quality-of-life concerns of African American breast cancer survivors within rural North Carolina: Blending the techniques of photovoice and grounded theory. *Qualitative Health Research, 15,* 99–115. doi: 10.1177/1049732304270766

Lord, C. (2004). *The summer of her baldness: A cancer improvisation.* Austin, TX: University of Texas Press.

Lorde, A. (1984). The master's tools will never dismantle the master's house. *Sister outsider: Essays and speeches* (pp. 110–114). Berkley, CA: Crossing Press.

Lorimer, H. (2005). Cultural geography: The busyness of being "more-than-representational". *Progress in Human Geography, 29,* 83–94.

Loveday, V. (2015). Embodying deficiency through 'affective practice': Shame, relationality, and the lived experience of social class and gender in higher education. *Sociology* (online). doi: 10.1177/0038038515589301. Retrieved from: http://research.gold.ac.uk/11583/

Lund, L. W., Schmiegelow, K., Rechnitzer, C., & Johansen, C. (2011). A systematic review on psychological late effects of childhood cancer: Structure of society and methodological pitfalls may change the conclusion. *Pediatric Blood Cancer, 56,* 532–543. doi: 10.1002/pbc.22883

Lupton, D. (2015). Quantified sex: A critical analysis of sexual and reproductive self-tracking using apps. *Culture, health & sexuality, 17,* 440–453. doi: 10.1080/13691058.2014.920528

Lyons, A. C., Emslie, C., & Hunt, K. (2014). Staying 'in the zone' but not passing the 'point of no return': Embodiment, gender and drinking in mid-life. *Sociology of Health & Illness, 36,* 264–277. doi: 10.1111/1467-9566.12103

MacLure, M. (2013). Classification or wonder? Coding as an analytic practice in qualitative research. In B. Coleman & J. Ringrose (Eds.), *Deleuze and research methodologies* (pp. 164–183). Edinburgh, UK: Edinburgh University Press.

Madison, D. S. (2005). *Critical ethnography: Method, ethics, and performance.* Thousand Oaks, CA: Sage.

Madsen, K. D. (2015). Research dissonance. *Geoforum, 65,* 192–200. doi: 10.1016/j.geoforum.2015.07.020

Mailman, J. B. (2012). Seven metaphors for (music) listening: DRAMaTIC. *Journal of Sonic Studies, 2*(1). Retrieved from: http://journal.sonicstudies.org/vol02/nr01/a03

Mairs, N. (1997). Carnal acts. In K. Conboy, N. Medina, & S. Stanbury (Eds.), *Writing on the body: Female embodiment and feminist theory* (pp. 296–305). New York: Columbia University Press.

Maiter, S., Simich, L., Jacobson, N., & Wise, J. (2008). Reciprocity: An ethic for community-based participatory action research. *Action Research, 6,* 305–325. doi: 10.1177/1476750307083720

Makagon, D., & Neumann, M. (2009). *Recording culture: Audio documentary and the ethnographic experience.* Thousand Oaks, CA: Sage.

Manning, E. (2013). *Always more than one: Individuation's dance.* Durham, NC: Duke University Press.

Manning, J. (2015a). Paradoxes of (im)purity: Affirming heteronormativity and queering heterosexuality in family discourses of purity pledges. *Women's Studies in Communication, 38,* 99–117. doi: 10.1080/07491409.2014.954687

Manning, J. (2015b). Positive and negative communicative behaviors in coming-out conversations. *Journal of Homosexuality, 62,* 67–97. doi: 10.1080/00918369.2014.957127

Manning, J., & Kunkel, A. (2014). Making meaning of meaning-making research: Using qualitative research for studies of social and personal relationships. *Journal of Social and Personal Relationships, 31,* 433–441. doi: 10.1177/0265407514525890

Markham, A. N. (2005). The methods, politics, and ethics of representation in online ethnography. In N. K. Denzin & Y. Lincoln (Eds.), *Handbook of qualitative research* (3rd ed. pp. 793–820). Thousand Oaks, CA: Sage.

Markham, A. N. (2013). Undermining 'data': A critical examination of a core term in scientific inquiry. *First Monday, 18*(10). Retrieved from: 10.5210/fm.v18i10.4868

Marshall, H. (1999). Our bodies, ourselves: Why we should add old fashioned empirical phenomenology to the new theories of the body. In J. Price & M. Shildrick (Eds.), *Feminist theory and the body: A reader* (pp. 64–75). New York: Routledge.

Martin, A. D., & Kamberelis, G. (2013). Mapping not tracing: Qualitative educational research with political teeth. *International Journal of Qualitative Studies in Education, 26,* 668–679. doi: 10.1080/09518398.2013.788756

Martinez, J. A. (2000). *Phenomenology of Chicana experience and identity: Communication and transformation in praxis.* New York: Rowman & Littlefield.

Mazzei, L. A., & McCoy, K. (2010). Thinking with Deleuze in qualitative research. *International Journal of Qualitative Studies in Education, 23,* 503–509. doi: 10.1080/09518398.2010.500634

McDonald, J. (2013). Coming out in the field: A queer reflexive account of shifting researcher identity. *Management Learning, 44,* 127–143. doi: 10.1177/1350507612473711

McRuer, R. (2004). Composing bodies; Or, de-composition: Queer theory, disability studies, and alternative corporealities. *JAC: A Quarterly Journal for the Interdisciplinary Study of Rhetoric, Culture, Literacy, and Politics, 24,* 47–76.

Meier, B. P., & Robinson, M. D. (2004). Why the sunny side is up. *Psychological Science, 15,* 243–247. doi: 10.1111/j.0956-7976.2004.00659.x

Merleau-Ponty, M. (1962). *Phenomenology of perception.* London, UK: Routledge.

Merleau-Ponty, M. (1968). *The visible and the invisible.* Evanston, IL: Northwestern University Press.

Mertens, D. M. (2007). Transformative paradigm mixed methods and social justice. *Journal of mixed methods research, 1,* 212–225. doi: 10.1177/1558689807302811

Mertens, D. M. (2010). Transformative mixed methods research. *Qualitative Inquiry, 16,* 469–474. doi: 10.1177/1077800410364612

Mertens, D. M. (2012). Transformative mixed methods: Addressing inequities. *American Behavioral Scientist, 56,* 802–813. doi: 10.1177/0002764211433797

Michel, A. (2015). Dualism at work: The social circulation of embodiment theories in use. *Signs and Society, 3,* S41–S69. doi: 10.1086/679306

Mies, M. (1983). Towards a methodology for feminist research. In G. Bowles & R. D. Klein (Eds.), *Theories of women's studies* (pp. 117–138). London, UK: Routledge.

Miller, K. I. (2000). Common ground from the post-positivist perspective: From "straw-person" argument to collaborative coexistence. In S. R. Corman & M. S. Poole (Eds.), *Perspectives on organizational communication: Finding common ground* (pp. 47–67). New York: Guilford.

Miller, K. (2014). *Organizational communication: Approaches and processes* (7th ed.). Belmont, CA: Wadsworth.

Miller-Day, M. (2008, May). Translational performances: Toward relevant, engaging, and empowering social science. In *Forum Qualitative Sozialforschung/Forum: Qualitative Social Research, 9*(2). Retrieved from: http://www.qualitative-research.net/index.php/fqs/article/viewArticle/402

Miller-Day, M., Conway, J. J., & Hecht, M. L. (2013). REAL adventures of a very strange day. Drug prevention comic. Los Angeles, CA: D.A.R.E. Catalog.

Minge, J. M. (2007). The stained body: A fusion of embodied art on rape and love. *Journal of Contemporary Ethnography, 36,* 252–280. doi: 10.1177/91241606287701

Minow, M. (1990). *Making all the difference: Inclusion, exclusion and the American law*. Ithaca, NY: Cornell University Press.

Mirza, H. S. (2009). Plotting a history: Black and postcolonial feminisms in 'new times'. *Race Ethnicity and Education, 12,* 1–10. doi: 10.1080/13613320802650899

Mishler, E. G. (1986). *Research interviewing: Context and narrative*. Cambridge, MA: Harvard University Press.

Mishler, E. G. (1991). Representing discourse: The rhetoric of transcription. *Journal of Narrative and Life History, 1,* 255–280.

Miyake, E. (2013). Understanding music and sexuality through ethnography: Dialogues between queer studies and music. *Transposition: Musique et Sciences Sociales, 3,* 2–17. Retrieved from: http://transposition.revues.org/150#quotation

Modaff, J. V., & Modaff, D. P. (2000). Technical notes on audio recording. *Research on Language and Social Interaction, 33,* 101–118. doi: 10.1207/S15327973RLSI3301_4

Mohanty, C. T. (1988). Under Western eyes: Feminist scholarship and colonial discourses. *Feminist Review, 30,* 61–88.

Moloney, S. (2011). Focus groups as transformative spiritual encounters. *International Journal of Qualitative Methods, 10,* 58–72. doi: 10.1177/160940691101000105

Montez, J. K., & Karner, T. X. (2005). Understanding the diabetic body-self. *Qualitative Health Research, 15,* 1086–1104. doi: 10.1177/104973230527667

Murphy, R. F. (2001). *The body silent*. New York: Norton.

Murray, C. D. (2004). An interpretative phenomenological analysis of the embodiment of artificial limbs. *Disability and Rehabilitation, 26,* 963–973. doi: 10.1080/09638280410001696764

Myers, W. B. (2008). Straight and white: Talking with my mouth full. *Qualitative Inquiry, 14,* 160–171. doi: 10.1177/1077800407308905

Myers, W. B., & Alexander, B. K. (2010). (Performance is) Metaphors as methodological tools in qualitative inquiry. *International Review of Qualitative Research, 3,* 163–172.

Nabbali, E. M. (2009). A 'mad' critique of the social model of disability. *International Journal of Diversity in Organisations, Communities, and Nations, 9*(4), 1–12.

Nabhan-Warren, K. (2011). Embodied research and writing: A case for phenomenologically oriented religious studies ethnographies. *Journal of the American Academy of Religion, 79,* 378–407. doi: 10.1093/jaarel/lfq079

Nafus, D., & Sherman, J. (2014). Big data, big questions. This one does not go up to 11: The quantified self movement as an alternative big data practice. *International Journal of Communication, 8,* 1784–1794. Retrieved from: http://ijoc.org/index.php/ijoc/article/view/2170

Natvik, E., Gjengedal, E., Moltu, C., & Råheim, M. (2014). Re-embodying eating: Patients' experiences 5 years after bariatric surgery. *Qualitative Health Research, 24,* 1700–1710. doi: 10.1177/1049732314548687

Nayar, P. K. (2014). *Posthumanism.* Malden, MA: Polity.

Neff, K. (2003). Self-compassion: An alternative conceptualization of a healthy attitude toward oneself. *Self and Identity, 2,* 85–101. doi: 10.1080/15298860309032

Nelson, L. D., & Simmons, J. P. (2009). On southbound ease and northbound fees: Literal consequences of the metaphoric link between vertical position and cardinal direction. *Journal of Marketing Research, 46,* 715–724. doi: 10.1509/jmkr.46.6.715

Nelson, S. (1998). Intersections of eros and ethnography. *Text and Performance Quarterly, 18,* 1–21. doi: 10.1080/10462939809366206

Newman, J. I. (2011). [Un]Comfortable in my own skin: Articulation, reflexivity, and the duality of self. *Cultural Studies ↔ Critical Methodologies, 11,* 545–557. doi: 10.1177/1532708611426110

Noddings, N. (1984). *Caring: A feminine approach to ethics & moral education.* Berkeley, CA: University of California Press.

Nordmarken, S. (2014). Becoming ever more monstrous: Feeling transgender. *Qualitative Inquiry, 20,* 37–50. doi: 10.1177/1077800413508531

Nordstrom, S. N. (2013). Object-interviews: Folding, unfolding, and refolding perceptions of objects. *International Journal of Qualitative Methods, 12,* 237–257. doi: 10.1177/160940691301200111

Nordstrom, S. N. (2015). Not so innocent anymore: Making recording devices matter in qualitative interviews. *Qualitative Inquiry, 21,* 388–401. doi: 10.1177/1077800414563804

Novak, D. R. (2010). Democratizing qualitative research: Photovoice and the study of human communication. *Communication Methods and Measures, 4,* 291–310. doi: 10.1080/19312458.2010.527870

Novick, G. (2008). Is there a bias against telephone interviews in qualitative research? *Research in Nursing & Health, 31,* 391–398. doi: 10.1002/nur.20259

Ntelioglou, B. Y. (2015). Embodied multimodality framework: Examining language and literacy practices of English language learners in drama classrooms. In M. Perry & C. L. Medina (Eds.), *Methodologies of embodiment: Inscribing bodies in qualitative research* (pp. 86–101). New York: Routledge.

Ochs, E. (1979). Transcription as theory. *Developmental Pragmatics, 10,* 43–72.

O'Connell, N. P. (2016). "Passing as normal": Living and coping with the stigma of Deafness. *Qualitative Inquiry, 22,* 651–661. doi: 1077800416634729

Olesen, V. (2011). Feminist qualitative research in the millennium's first decade. In N. K. Denzin & Y. S. Lincoln (Eds.). *The Sage handbook of qualitative research* (4th ed., pp. 129–146). Thousand Oaks, CA: Sage.

Onwuegbuzie, A. J., & Byers, V. T. (2014). An exemplar for combining the collection, analysis, and interpretations of verbal and nonverbal data in qualitative research. *International Journal of Education, 6,* 183–246. doi: 10.5296/ije.v6i1.4399

Onwuegbuzie, A. J., Leech, N. L., & Collins, K. M. (2008). Interviewing the interpretive researcher: A method for addressing the crises of representation, legitimation, and praxis. *International Journal of Qualitative Methods, 7*(4), 1–18. doi: 10.1177/160940690800700401

Pande, A. (2010). Commercial surrogacy in India: Manufacturing a perfect mother-worker. *Signs, 35,* 969–992. doi: 10.1086/651043

Parker Webster, J., & Marques da Silva, S. (2013). Doing educational ethnography in an online world: Methodological challenges, choices and innovations. *Ethnography and Education, 8,* 123–130. doi: 10.1080/17457823.2013.792508

Parry, D. C. (2006). Women's lived experiences with pregnancy and midwifery in a medicalized and fetocentric context: Six short stories. *Qualitative Inquiry, 12,* 459–471. doi: 10.1177/1077800406286225

Paterson, M. (2009). Haptic geographies: Ethnography, haptic knowledges and sensuous dispositions. *Progress in Human Geography, 33,* 766–788. doi: 10.1177/0309132509103155

Patti, C. J. (2015). Sharing "a big kettle of soup": Compassionate listening with a Holocaust survivor. In S. High (Ed.), *Beyond testimony and trauma: Oral history in the aftermath of mass violence* (pp. 192–211). Vancouver, BC: University of British Columbia Press.

Pausé, C. (2014). X-static process: Intersectionality within the field of fat studies. *Fat Studies, 3,* 80–85. doi: 10.1080/21604851.2014.889487

Pazzaglia, M., & Molinari, M. (2016). The embodiment of assistive devices—from wheelchair to exoskeleton. *Physics of Life Reviews, 16,* 163–175. doi: 10.1016/j.plrev.2015.11.006

Peluso, N. M. (2011). "Cruisin' for a bruisin'": Women's flat track roller derby. In C. Bobel & S. Kwan (Eds.), *Embodied resistance: Challenging the norms, breaking the rules* (pp. 37–47). Nashville, TN: Vanderbilt University Press.

Peralta, R. L. (2007). College alcohol use and the embodiment of hegemonic masculinity among European American men. *Sex Roles, 56,* 741–756. doi: 10.1007/s11199-007-9233-1

Perrier, M. J., & Kirkby, J. (2013). Taming the 'Dragon': Using voice recognition software for transcription in disability research within sport and exercise psychology. *Qualitative Research in Sport, Exercise and Health, 5,* 103–108. http://dx.doi.org/10.108 0/2159676X.2012.712996

Perry, M., & Medina, C. (2011). Embodiment and performance in pedagogy research: Investigating the possibility of the body in curriculum experience. *Journal of Curriculum Theory, 27*(3), 62–75. Retrieved from: http://search.proquest.com/docview/1010056066?pq-origsite=gscholar

Perry, M., & Medina, C. L. (2015). Introduction: Working through the contradictory terrain of the body in qualitative research. In M. Perry & C. L. Medina (Eds.), *Methodologies of embodiment: Inscribing bodies in qualitative research* (pp. 1–13). New York: Routledge.

Pettinger, L. (2005). Gendered work meets gendered goods: Selling and service in clothing retail. *Gender, Work & Organization, 12,* 460–478. doi: 10.1111/j.1468-0432.2005.00284.x

Phoenix, C., & Sparkes, A. C. (2007). Sporting bodies, ageing, narrative mapping and young team athletes: An analysis of possible selves. *Sport, Education and Society, 12,* 1–17. doi: 10.1080/13573320601081468

Pilkey, B. (2013). LGBT homemaking in London, UK: The embodiment of mobile homemaking imaginaries. *Geographical Research, 51,* 159–165. doi: 10.1111/j.1745-5871.2012.00752.x

Pink, S. (2005). Dirty laundry: Everyday practice, sensory engagement and the constitution of identity. *Social Anthropology, 13,* 275–290. doi: 10.1017/S0964028205001540

Pink, S. (2007). The sensory home as a site of consumption: Everyday laundry practices and the production of gender. In E. Casey & L. Martens (Eds.), *Gender and consumption: Domestic cultures and the commercialisation of everyday life* (pp. 163–180). London and New York: Routledge.

Pink, S. (2009). Urban social movements and small places: Slow cities as sites of activism. *City, 13,* 451–465. doi: 10.1080/13604810903298557

Pink, S. (2011). Multimodality, multisensoriality and ethnographic knowing: Social semiotics and the phenomenology of perception. *Qualitative Research, 11,* 261–276. doi: 10.1177/1468794111399835

Pink, S. (2015). *Doing sensory ethnography* (2nd ed.). Thousand Oaks, CA: Sage.

Pink, S., Mackley, K. L., & Moroşanu, R. (2015). Hanging out at home: Laundry as a thread and texture of everyday life. *International Journal of Cultural Studies, 18,* 209–224. doi: 10.1177/1367877913508461

Pitts, V. (2003). *In the flesh: The cultural politics of body modification.* New York: Palgrave Macmillan.

Pitts-Taylor, V. (2007). *Surgery junkies: Wellness and pathology in cosmetic culture.* New Brunswick, NJ : Rutgers University Press.

Pitts-Taylor, V. (2015). A feminist carnal sociology?: Embodiment in sociology, feminism, and naturalized philosophy. *Qualitative Sociology, 38,* 19–25. doi: 10.1007/s11133-014-9298-4

Pitts-Taylor, V. (2016). *The brain's body: Neuroscience and corporeal politics.* Durham, NC: Duke University Press.

Potter, W. J. (1996). *An analysis of thinking and research about qualitative methods.* Mahwah, NJ: Erlbaum.

Powell, K. M., & Takayoshi, P. (2003). Accepting roles created for us: The ethics of reciprocity. *College Composition and Communication, 54,* 394–422. doi: 10.2307/3594171

Preissle, J. (2007). Feminist research ethics. In S. N. Hesse-Biber (Ed.), *Handbook of feminist research: Theory and praxis* (pp. 515–532). Thousand Oaks, CA: Sage.

Priya, K. R. (2010). The research relationship as a facilitator of remoralization and self-growth: Postearthquake suffering and healing. *Qualitative Health Research, 20,* 479–495. doi: 10.1177/1049732309360419

Probyn, E. (1991). This body which is not one: Speaking an embodied self. *Hypatia, 6,* 111–124. doi: 10.1111/j.1527-2001.1991.tb00258.x

Probyn, E. (2011). Glass selves: Emotions, subjectivity, and the research process. In S. Gallagher (Ed.), *The Oxford handbook of the self* (pp. 681–695). Oxford, UK: Oxford University Press.

Quinlan, M. M. (2010). Dancing wheels: Integration and diversity. In L. Black (Ed.), *Group communication: Cases for analysis, appreciation, and application* (pp. 43–48). Dubuque, IA: Kendall Hunt Publishing Company.

Quinlan, M. M., & Bates, B. R. (2008). Dances and discourses of (dis) ability: Heather Mills's embodiment of disability on Dancing with the Stars. *Text and Performance Quarterly, 28,* 64–80. doi: 10.1080/10462930701754325

Quinlan, M. M., & Bates, B. R. (2014). Unsmoothing the cyborg: Technology and the body in integrated dance. *Disability Studies Quarterly, 34*(4). Retrieved from: http://dsq-sds.org/article/view/3783/3792

Quinlan, M. M., & Harter, L. M. (2010). Meaning in motion: The embodied poetics and politics of Dancing Wheels. *Text and Performance Quarterly, 30,* 374–395. doi: 10.1080/10462937.2010.510911

Race, K. (2012). 'Frequent sipping': Bottled water, the will to health and the subject of hydration. *Body & Society, 18*(3&4), 72–98. doi: 10.1177/1357034X12450592

Rager, K. B. (2005). Self-care and the qualitative researcher: When collecting data can break your heart. *Educational Researcher, 34*(4), 23–27. doi: 10.3102/0013189x034004023

Rambo Ronai, C. (1995). Multiple reflections on childhood sex abuse: An argument for a layered account. *Journal of Contemporary Ethnography, 23,* 395–426. doi: 10.1177/089124195023004001

Råsmark, G., Richt, B., & Rudebeck, C. E. (2014). Touch and relate: Body experience among staff in habilitation services. *International Journal of Qualitative Studies on Health and Well-being, 9*. doi:10.3402/qhw.v9.21901

Rattine-Flaherty, E. (2014). Participatory sketching as a tool to address student's public speaking anxiety. *Communication Teacher, 28*, 26–31. doi:10.1080/17404622.2013.839048

Rawlins, W. K. (2007). Living scholarship: A field report. *Communication Methods and Measures, 1*, 55–63. doi: 10.1080/19312450709336662

Rawlins, W. K. (2013). Sample. *Liminalities, 9*(3). Retrieved: http://liminalities.net/9-3/sample.html.

Razon, N., & Ross, K. (2012). Negotiating fluid identities: Alliance-building in qualitative interviews. *Qualitative Inquiry, 18*, 494–503. doi: 10.1177/1077800412442816

Reagon, B. J. (1983). Coalition politics: Turning the century. In B. Smith (Ed.), *Home girls: A Black feminist anthology* (pp. 343–355). New Brunswick, NJ: Rutgers University Press.

Reay, D. (1997). Feminist theory, habitus, and social class: Disrupting notions of classlessness. *Women's Studies International Forum, 20*, 225–233. doi: 10.1016/S0277-5395(97)00003-4

Redford, P. & Sorkin, A. (2001, March 27). Somebody's going to emergency, somebody's going to jail [Television series episode]. In J. Yu (Director), *The West Wing*. New York: National Broadcasting Company.

Reichertz, J. (2004). Abduction, deduction and induction in qualitative research. In U. Flick, E. von Kardoff, & I. Steinke (Eds.). *A companion to qualitative research* (pp. 159–163). Thousand Oaks, CA: Sage.

Reilly, R. C. (2013). Found poems, member checking and crises of representation. *The Qualitative Report, 18*(30), 1–18. Retrieved from: http://www.nova.edu/ssss/QR/QR18/reilly30.pdf

Reinharz, S. (1992). *Feminist methods in social research.* New York: Oxford University Press.

Rennie, D. L., & Fergus, K. D. (2006). Embodied categorizing in the grounded theory method: Methodical hermeneutics in action. *Theory & Psychology, 16*, 483–503. doi: 10.1177/0959354306066202

Riach, K., & Warren, S. (2015). Smell organization: Bodies and corporeal porosity in office work. *Human Relations, 68*, 789–809. doi: 0018726714545387.

Rice, C. (2015). Rethinking fat from bio- to body-becoming pedagogies. *Cultural Studies ↔ Critical Methodologies, 15*, 387–397. doi: 10.1177/1532708615611720

Rice, T. (2010). Learning to listen: Auscultation and the transmission of auditory knowledge. *Journal of the Royal Anthropological Institute 16*, S41–S61. doi: 10.1111/j.1467-9655.2010.01609.x

Richards, R. (2008). Writing the othered self: Autoethnography and the problem of objectification in writing about illness and disability. *Qualitative Health Research, 18*, 1717–1728. doi: 10.1177/1049732308325866

Richardson, A. & Cherry, A. (2011). Anorexia as a choice: Constructing a new community of health and beauty through pro-ana websites. In C. Bobel & S. Kwan (Eds.), *Embodied resistance: Challenging the norms, breaking the rules* (pp. 119–129). Nashville, TN: Vanderbilt University Press.

Richardson, L. (1992). The consequences of poetic representation: Writing the other, rewriting the self. In C. Ellis & M. G. Flaherty (Eds.), *Investigating subjectivity: Research on lived experience* (pp. 125–140). Thousand Oaks, CA: Sage.

Richardson, L. (2000). Writing: A method of inquiry. In N. K. Denzin & Y. S. Lincoln (Eds.), *Handbook of qualitative research* (2nd ed., pp. 923–943). Thousand Oaks, CA: Sage.

Riessman, C. K. (1987). When gender is not enough: Women interviewing women. *Gender & Society, 1,* 172–207. doi: 10.1177/0891243287001002004

Riley, M. (2010). Emplacing the research encounter: Exploring farm life histories. *Qualitative Inquiry, 16,* 651–662. doi: 10.1177/1077800410374029

Ringrose, J., & Renold, E. (2014). "F**k rape!": Exploring affective intensities in a feminist research assemblage. *Qualitative Inquiry, 20,* 772–780. doi: 10.1177/1077800414530261

Ritenburg, H., Young Leon, A. E., Linds, W., Nadeau, D. M., Goulet, L. M., Kovach, M., & Marshall, M. (2014). Embodying decolonization: Methodologies and indigenization. *AlterNative: An International Journal of Indigenous Peoples, 10(1),* 68–80. Retrieved from: http://www.content.alternative.ac.nz/index.php/alternative/article/view/294

Roberts, R. A. (2013). How do we quote black and brown bodies? Critical reflections on theorizing and analyzing embodiments. *Qualitative Inquiry, 19,* 280–287. doi: 10.1177/1077800412471512

Rogers, R. (2003). *A critical discourse analysis of family literacy practices: Power in and out of print.* Mahwah, NJ: Erlbaum.

Roof, J. (2007). Authority and representation in feminist research. In S. N. Hesse-Biber (Ed.), *Handbook of feminist research: Theory and praxis* (pp. 425–442). Thousand Oaks, CA: Sage.

Rosaldo, M. Z. (1984). Toward an anthropology of self and feeling. In R. A. Shweder & R. A. LeVine (Eds.), *Culture theory: Essays on mind, self, and emotion* (pp. 137–157). Cambridge, UK: Cambridge University Press.

Rose, G. (1999). Women and everyday spaces. In J. Price & M. Shildrick (Eds.), *Feminist theory and the body: A reader* (pp. 359–370). New York: Routledge.

Rosenberg, R., & Oswin, N. (2015). Trans embodiment in carceral space: Hypermasculinity and the US prison industrial complex. *Gender, Place & Culture, 22,* 1269–1286. doi: http://dx.doi.org/10.1080/0966369X.2014.969685

Rothman, B. K. (1986). Reflections: On hard work. *Qualitative Sociology, 9,* 48–53. doi: 10.1007/BF00988248

Ryan, K., Todres, L., & Alexander, J. (2011). Calling, permission, and fulfillment: The interembodied experience of breastfeeding. *Qualitative Health Research, 21,* 731–742. doi: 10.1177/1049732310392591

Saldaña, J. (2003). Dramatizing data: A primer. *Qualitative Inquiry, 9,* 218–236. doi: 10.1177/1077800402250932

Saldaña, J. (2005). *Ethnodrama: An anthology of reality theatre.* New York: Rowman.

Saldaña, J. (2011). *Ethnotheatre: Research from page to stage.* Walnut Creek, CA: Left Coast Press.

Saldaña, J. (2014). Blue-collar qualitative research: A rant. *Qualitative Inquiry, 20,* 976–980. doi: 10.1177/1077800413513739

Sandelowski, M. (2002). Reembodying qualitative inquiry. *Qualitative Health Research, 12,* 104–115. doi: 10.1177/1049732302012001008

Sanon, M. A., Evans-Agnew, R. A., & Boutain, D. M. (2014). An exploration of social justice intent in photovoice research studies from 2008 to 2013. *Nursing Inquiry, 21,* 212–226. doi: 10.1111/nin.12064

Satinsky, S., & Ingraham, N. (2014). At the intersection of public health and fat studies: Critical perspectives on the measurement of body size. *Fat Studies, 3,* 143–154. doi: 10.1080/21604851.2014.889505

Scarry, E. (1985). *The body in pain: The making and unmaking of the world.* New York: Oxford University Press.

Schatzki, T. R. (2001). Introduction: Practice theory. In T. R. Schatzki, K. Knorr-Cetina, & E. Von Savigny (Eds.). *The practice turn in contemporary theory* (pp. 1–14). Hove, UK: Psychology Press.

Schipper, K., Abma, T. A., van Zadelhoff, E., van de Griendt, J., Nierse, C., & Widdershoven, G. A. (2010). What does it mean to be a patient research partner? An ethnodrama. *Qualitative Inquiry, 16,* 501–510. doi: 10.1177/1077800410364351

Schmitz, S., & Höppner, G. (2014). Neurofeminism and feminist neurosciences: A critical review of contemporary brain research. *Frontiers in Human Neuroscience, 8,* Article 546. doi: 10.3389/fnhum.2014.00546

Scott, J. (2012). Problematizing a researcher's performance of "insider status": An autoethnography of "designer disabled" identity. *Qualitative Inquiry, 19,* 101–115. doi: 10.1177/1077800412462990

Scott, J. A. (2015). Almost passing: A performance analysis of personal narratives of physically disabled femininity. *Women's Studies in Communication, 38,* 227–249. doi: 10.1080/07491409.2015.1027023

Scott, K. D. (2013). Communication strategies across cultural borders: Dispelling stereotypes, performing competence, and redefining black womanhood. *Women's Studies in Communication, 36,* 312–329. doi: /10.1080/07491409.2013.831005

Sedgwick, E. K. (1993). *Tendencies.* Durham, NC: Duke University Press.

Sekimoto, S. (2012). A multimodal approach to identity: Theorizing the self through embodiment, spatiality, and temporality. *Journal of International and Intercultural Communication, 5,* 226–243. doi: 10.1080/17513057.2012.689314

Seymour, W. (2007). Exhuming the body: Revisiting the role of the visible body in ethnographic research. *Qualitative Health Research, 17,* 1188–1197. doi: 10.1177/1049732307308517

Sharma, S., Reimer-Kirkham, S., & Cochrane, M. (2009). Practicing the awareness of embodiment in qualitative health research: Methodological reflections. *Qualitative Health Research, 19,* 1642–1650. doi: 10.1177/1049732309350684

Sheridan, J., Chamberlain, K., & Dupuis, A. (2011). Timelining: Visualizing experience. *Qualitative Research, 11,* 552–569. doi: 10.1177/1468794111413235

Shildrick, M. (2013). Re-imagining embodiment: Prostheses, supplements and boundaries. *Somatechnics, 3,* 270–286. doi: 10.3366/soma.2013.0098

Shildrick, M. (2015). "Why should our bodies end at the skin?": Embodiment, boundaries, and somatechnics. *Hypatia, 30,* 13–29. doi: 10.1111/hypa.12114

Shilling, C. (2012). *The body and social theory* (3rd ed.). Thousand Oaks, CA: Sage.

Sholock, A. (2012). Methodology of the privileged: White anti-racist feminism, systematic ignorance, and epistemic uncertainty. *Hypatia, 27,* 701–714. doi: 10.1111/j.1527-2001.2012.01275.x

Shreve, A. (2010). Origins of "Sea Glass." Retrieved from: http://www.fictionaddiction.net/Behind-the-Books/origins-of-sea-glass-by-anita-shreve/Page-2-alt=-The-Idea-Takes-Shape.html

Shuy, R. W. (2003). In-person versus telephone interviewing. In J. A. Holstein & J. F. Gubrium (Eds.), *Inside interviewing: New lenses, new concerns* (pp. 175–193). Thousand Oaks, CA: Sage.

Simmonds, F. N. (1999). My body, myself: How does a black woman do sociology? In J. Price & M. Shildrick (Eds.), *Feminist theory and the body: A reader* (pp. 50–63). New York: Routledge.

Simonds, V. W., & Christopher, S. (2013). Adapting Western research methods to indigenous ways of knowing. *American Journal of Public Health, 103*, 2185–2192. doi: 10.2105/AJPH.2012.301157

Simpson, A., & Quigley, C. F. (2016). Member checking process with adolescent students: Not just reading a transcript. *The Qualitative Report, 21*(2), 377–392. Retrieved from http://nsuworks.nova.edu/tqr/vol21/iss2/12

Singhal, A., & Rattine-Flaherty, E. (2006). Pencils and photos as tools of communicative research and praxis: Analyzing Minga Perú's quest for social justice in the Amazon. *International Communication Gazette, 68*, 313–330. doi: 10.1177/1748048506065764

Singhal, A., Harter, L. M., Chitnis, K., & Sharma, D. (2007). Participatory photography as theory, method and praxis: Analyzing an entertainment-education project in India. *Critical Arts: A Journal of South-North Cultural Studies, 21*, 212–227. doi: 10.1080/02560040701398897

Slatman, J., Halsema, A., & Meershoek, A. (2015). Responding to scars after breast surgery. *Qualitative Health Research, 26*, 1614–1626. doi: 1049732315591146

Smith, L., & Fershleiser, R. (2008). *Not quite what I was planning: Six-word memoirs by writers famous and obscure* (revised ed.). New York: HarperCollins.

Smith, V. J. (2009). Ethical and effective ethnographic research methods: A case study with Afghan refugees in California. *Journal of Empirical Research on Human Research Ethics, 4*(3), 59–72. doi: 10.1525/jer.2009.4.3.59

Smith-Nonini, S. (2011). The illegal and the dead: Are Mexicans renewable energy? *Medical Anthropology, 30*, 454–474. doi: 10.1080/01459740.2011.577045

Sobchack, V. (2010). Living a 'phantom limb': On the phenomenology of bodily integrity. *Body & Society, 16*(3), 51–67. doi: 10.1177/1357034X10373407

Sotirin, P. (2010). Autoethnographic mother-writing: Advocating radical specificity. *Journal of Research Practice, 6*(1). Article M9. Retrieved from: http://jrp.icaap.org/index.php/jrp/article/view/220/220

Sotirin, P. J. (2016). Silencers: Governmentality, gender, and the ban on gun violence research. In C. Squires (Ed.), *Dangerous discourses: Feminism, guns, violence, & civic life* (pp. 27–53). New York: Peter Lang.

Sotirin, P. J., & Ellingson, L. L. (2013). *Where the aunts are: Family, feminism, and kinship in popular culture*. Waco, TX: Baylor University Press.

Sparkes, A. C. (2012). Ageing and embodied masculinities in physical activity settings. In E. Tulle & C. Phoenix (Eds.), *Physical activity and sport in later life: Critical perspectives* (pp. 137). London, UK: Palgrave Macmillan.

Sparkes, A. C. (2015). Ageing and embodied masculinities in physical activity settings: From flesh to theory and back again. In E. Tulle & C. Phoenix (Eds.), *Physical activity and sport in later life* (pp. 137–148). London, UK: Palgrave Macmillan.

Sparkes, A. C., & Smith, B. (2002). Sport, spinal cord injury, embodied masculinities, and the dilemmas of narrative identity. *Men and Masculinities, 4*, 258–285. doi: 10.1177/1097184X02004003003

Sparkes, A. C., & Smith, B. (2008). Men, spinal cord injury, memories and the narrative performance of pain. *Disability & Society, 23*, 679–690. doi: 10.1080/09687590802469172

Spry, T. (2001). Performing autoethnography: An embodied methodological praxis. *Qualitative Inquiry, 7*, 706–732. doi: 10.1177/107780040100700605

Stelter, R. (2010). Experience-based, body-anchored qualitative research interviewing. *Qualitative Health Research, 20*, 859–867. doi: 10.1177/1049732310364624

Stewart, H., Gapp, R. & Harwood, I. (2017). Exploring the alchemy of qualitative management research: Seeking trustworthiness, credibility and rigor through crystallization. *The Qualitative Report, 22*, (1–19). Retrieved from: http://eprints.soton.ac.uk/401432/

Stodulka, T. (2015). Emotion work, ethnography, and survival strategies on the streets of Yogyakarta. *Medical Anthropology, 34*, 84–97. doi: 10.1080/01459740.2014.916706

St. Pierre, E. A., & Jackson, A. Y. (2014). Qualitative data analysis after coding. *Qualitative Inquiry, 20*, 715–719. doi: 10.1177/1077800414532435

St. Pierre, J. (2015). Cripping communication: Speech, disability, and exclusion in liberal humanist and posthumanist discourse. *Communication Theory, 25*, 330–348. doi: 10.1111/comt.12054

Strasser, D. S. (2016). 'You might want to call your father': An autoethnographic account of masculinity, relationships, and my father. *Journal of Family Communication, 16*, 64–75. doi: 10.1080/15267431.2015.1111214

Strean, W. B. (2011). Creating student engagement? HMM: Teaching and learning with humor, music, and movement. *Creative Education, 2*, 189–192. doi: 10.4236/ce.2011.23026

Swartz, S. (2011). 'Going deep' and 'giving back': Strategies for exceeding ethical expectations when researching amongst vulnerable youth. *Qualitative Research, 11*, 47–68. doi: 10.1177/1468794110385885

Tanner, C. A. (2006). Thinking like a nurse: A research-based model of clinical judgment in nursing. *Journal of Nursing Education, 45*, 204–211.

Thompson, M. (2010). The wisdom of vulnerability: A post-structural feminist exploration of healing in the aftermath of war. (Unpublished dissertation). Ohio University. Retrieved from: http://etd.ohiolink.edu/sendpdf.cgi/Thompson%20Marie.pdf?ohiou1276786036

Thomson, R. G. (1997). *Extraordinary bodies: Figuring physical disability in American culture and literature*. Columbia University Press.

Thoresen, L., & Öhlén, J. (2015). Lived observations: Linking the researcher's personal experiences to knowledge development. *Qualitative Health Research, 25*, 1589–1598. doi: 10.1177/1049732315573011

Thorp, L. (2006). *Pull of the earth: Participatory ethnography in the school garden*. Walnut Creek, CA: AltaMira Press.

Tillmann-Healy, L. M. (1996). A secret life in a culture of thinness: Reflections on body, food, and bulimia. In C. Ellis & A. P. Bochner (Eds.), *Composing ethnography: Alternative forms of qualitative writing* (pp. 76–108). Walnut Creek, CA: AltaMira Press.

Tracy, S. J. (2010). Qualitative quality: Eight "big-tent" criteria for excellent qualitative research. *Qualitative Inquiry, 16*, 837–851. doi: 10.1177/1077800410383121

Tracy, S. J. (2013). *Qualitative research methods: Collecting evidence, crafting analysis, communicating impact*. New York: Wiley.

Tracy, S. J., & Malvini Redden, S. (2016). Markers, metaphors, and meaning: Drawings as a visual and creative qualitative research methodology in organizations. In K. D. Elsbach and R. M. Kramer (Eds.), *Handbook of qualitative organizational research: Innovative pathways and ideas* (pp. 238–248). New York: Routledge.

Trier-Bieniek, A. (2012). Framing the telephone interview as a participant-centred tool for qualitative research: A methodological discussion. *Qualitative Research, 12*, 630–644. doi: 10.1177/1468794112439005

Trinh, T. M. (1999). Write your body and The body in theory. In J. Price & M. Shildrick (Eds.). *Feminist theory and the body. A reader* (pp. 258–266). New York: Routledge.

Turner, B. S. (2012). Introduction: The turn of the body. In B. S. Turner (Ed.), *Routledge handbook of body studies* (pp. 1–17). New York: Routledge.

Turner, P. K. (2002). Is childbirth with midwives natural? The gaze of the feminine and the pull of the masculine. *Qualitative Inquiry, 8,* 652–659. doi: 10.1177/107780002237016

Turner, P. K. (2004). Mainstreaming alternative medicine: Doing midwifery at the intersection. *Qualitative Health Research, 14,* 644–662. doi: 10.1177/1049732304263656

Turner, P. K., & Norwood, K. M. (2013). Body of research: Impetus, instrument, and impediment. *Qualitative Inquiry, 19,* 696–711. doi 10.1177/1077800413500928

Van Enk, A. A. (2009). The shaping effects of the conversational interview: An examination using Bakhtin's theory of genre. *Qualitative Inquiry, 15,* 1265–1286. doi: 10.1177/1077800409338029

Van Maanen, J. (2011). *Tales of the field: On writing ethnography* (2nd ed.). Chicago, IL: University of Chicago Press.

Van Manen, M. (1997). *Researching lived experience: Human science for an action sensitive pedagogy* (2nd ed.). London, UK: Althouse.

Vannini, P. (Ed.). (2015). *Non-representational methodologies: Re-envisioning research.* New York: Routledge.

Vannini, P., Ahluwalia-Lopez, G., Waskul, D., & Gottschalk, S. (2010). Performing taste at wine festivals: A somatic layered account of material culture. *Qualitative Inquiry, 16,* 378–396. doi: 10.1177/1077800410366939

Vitellone, N. (2010). Just another night in the shooting gallery?: The syringe, space, and affect. *Environment and Planning D: Society and Space, 28,* 867–880. doi: 10.1068/d12609

Wacquant, L. (2009). The body, the ghetto and the penal state. *Qualitative Sociology, 32,* 101–129. doi: 10.1007/s11133-008-9112-2

Wagner, P. E., Ellingson, L. L., & Kunkel, A. (2016). Pictures, patience, and practicalities: Lessons learned from using photovoice in applied communication contexts. *Journal of Applied Communication Research, 44,* 336–342. doi: 10.1080/00909882.2016.1192292

Walsh, D. J. (2010). Childbirth embodiment: Problematic aspects of current understandings. *Sociology of Health & Illness, 32,* 486–501. doi: 10.1111/j.1467-9566.2009.01207.x

Walton, J. (2015). Feeling it: Understanding Korean adoptees' experiences of embodied identity. *Journal of Intercultural Studies, 36,* 395–412. doi: 10.1080/07256868.2015.1049985

Wang, C. C. (1999). Photovoice: A participatory action research strategy applied to women's health. *Journal of Women's Health, 8,* 185–192. doi: 10.1089/jwh.1999.8.185

Wang, J. Y., Ramberg, R., & Kuoppala, H. (2012). User participatory sketching: A complementary approach to gather user requirements. *Proceedings of APCHI 2012: The 10th Asia Pacific Conference on Computer Human Interaction, 2,* 481–490. Retrieved from: http://mobile-life.org/sites/default/files/User%20Participatory%20Sketching-%20A%20Complementary%20Approach%20to%20Gather%20User%20Requirements.pdf

Warin, M. J., & Gunson, J. S. (2013). The weight of the word: Knowing silences in obesity research. *Qualitative Health Research, 23,* 1686–1696. doi: 10.1177/1049732313509894

Warr, D. J. (2004). Stories in the flesh and voices in the head: Reflections on the context and impact of research with disadvantaged populations. *Qualitative Health Research, 14,* 578–587. doi: 10.1177/1049732303260449

Warren, C. A. B., & Karner, T. X. (2009). *Discovering qualitative methods: Field research, interviews, and analysis* (2nd ed.). New York: Oxford University Press.

Waskul, D. D., & Van der Riet, P. (2002). The abject embodiment of cancer patients: Dignity, selfhood, and the grotesque body. *Symbolic Interaction, 25,* 487–513. doi: 10.1525/si.2002.25.4.487

Waskul, D. D., & Vannini, P. (2008). Smell, odor, and somatic work: Sense-making and sensory management. *Social Psychology Quarterly, 71,* 53–71.

Way, D., & Tracy, S. J. (2012). Conceptualizing compassion as recognizing, relating and (re) acting: A qualitative study of compassionate communication at hospice. *Communication Monographs, 79,* 292–315. doi: 10.1080/03637751.2012.697630

Weiss, G. (1999). *Body images: Embodiment as intercorporeality.* New York: Routledge.

West, C. (2001). *Race matters* (2nd ed.). New York: Vintage.

White, K. (2012). The body, social inequality and health. In B. S. Turner (Ed.), *The Routledge handbook of the body* (pp. 264–274). New York: Routledge.

Wiggins, S. (2002). Talking with your mouth full: Gustatory mmms and the embodiment of pleasure. *Research on Language and Social Interaction, 35,* 311–336. doi: 10.1207/ S15327973RLSI3503_3

Williams, C. J., Weinberg, M. S., & Rosenberger, J. G. (2013). Trans men: Embodiments, identities, and sexualities. *Sociological Forum, 38,* 719–741. doi: 10.1111/socf.12056

Willis, P. (2002). Don't call it poetry. *Indo-Pacific Journal of Phenomenology, 2*(1), 1–14. doi: 10.1080/20797222.2002.11433869.

Wilson, S. (2008). *Research is ceremony: Indigenous research methods.* Black Point, NS: Fernwood.

Wittgenstein, L. (1953). *Philosophical investigations.* Malden, MA: Blackwell.

Wood, J. (2012). *Gendered lives: Communication, gender, and culture* (9th ed.). Belmont, CA: Wadsworth.

Woodby, L. L., Williams, B. R., Wittich, A. R., & Burgio, K. L. (2011). Expanding the notion of researcher distress: The cumulative effects of coding. *Qualitative Health Research, 21,* 830–838. doi: 10.1177/1049732311402095

Woolf, V. (1989). *A room of one's own.* New York: Harcourt Brace.

Wray, N., Markovic, M., & Manderson, L. (2007). "Researcher saturation": The impact of data triangulation and intensive-research practices on the researcher and qualitative research process. *Qualitative Health Research, 17,* 1392–1402. doi: 10.1177/1049732307308308

Wright, K. O. (2009). *Putting 'sugar diabetes' on the table: Evaluating "The Sugar Plays" as entertainment-education in Appalachia.* Unpublished doctoral dissertation, Ohio University. Retrieved from: https://etd.ohiolink.edu/pg_10?0::NO:10:P10_ACCESSION_NUM: ohiou1245441476

Yamasaki, J., Geist-Martin, P. & Sharf, B. F. (2016). *Storied health and illness: Communicating personal, cultural, and political complexities.* Long Grove, IL: Waveland.

INDEX